Strategies for High-Tech Firms

Strategies for High-Tech Firms

Marketing, Economic, and Legal Issues

P.M. Rao • Joseph A. Klein

Routledge
Taylor & Francis Group
LONDON AND NEW YORK

To the memory of my parents and my brother
And to Sri Janardana Rao, who with Vasanta, made all the difference
Also for Shobha, Ramu, Nitya, and Pooja, with love
—P.M. Rao

To the memory of my wife, Roberta Klein
—Joseph A. Klein

First published 2013 by M.E. Sharpe

Published 2015 by Routledge
2 Park Square, Milton Park, Abingdon, Oxon OX14 4RN
711 Third Avenue, New York, NY 10017, USA

Routledge is an imprint of the Taylor & Francis Group, an informa business

Library of Congress Cataloging-in-Publication Data

Rao, P. M., 1938–
Strategies for high-tech firms : marketing, economic, and legal approaches / by P.M. Rao and Joseph A. Klein.
p. cm.
Includes bibliographical references and index.
ISBN 978-0-7656-1797-2 (hardcover : alk. paper)—ISBN 978-0-7656-1798-9 (pbk. : alk. paper)
1. High technology industries. 2. High technology industries—Marketing. 3. Technological innovations. 4. High technology industries—Case studies. 5. Technological innovations—Case studies. I. Klein, Joseph A. II. Title.

HC79.H53R363 2013
658.4′012—dc23 2012035533

ISBN 13: 9780765617989 (pbk)
ISBN 13: 9780765617972 (hbk)

Contents

Preface

Our motivation to write this book had its origins in our collective prior experience conducting and managing studies about the relationship between structure and innovative performance in the U.S. telecommunications industry. In this context, we were greatly inspired by the ideas and works of William D. Nordhaus, Sterling Professor of Economics at Yale University, with whom we worked briefly. In our subsequent research we began to examine, more broadly, certain unique problems faced by high-tech firms, such as the problem of *appropriability*—the difficulty firms face in fully capturing the benefits of their innovative efforts—from a strategy perspective. Our focus is on marketing strategy because of its central role in mitigating appropriability and other problems associated with marketing of high-tech products and services. We draw upon relevant concepts and frameworks from the industrial organization branch of economics and strategy literature. High-tech firms create intellectual property, which needs to be protected from infringement through legal rights in the form of patents and copyrights. Therefore, understanding intellectual property rights (IPRs) is essential to devising effective marketing strategies.

The content of this book differs from typical books on marketing strategy in several respects. First, our focus is on the unique problems faced by high-tech firms and solutions to such problems. Second, the book integrates relevant concepts from economics and law. Third, we consider, in a novel way, externalities associated with high-tech firms and explain their relationship to ethical issues, brand name, and reputation. Fourth, the book is more research-based than a typical textbook on this subject in the sense that it draws upon empirical evidence from scholarly journals to a greater extent. Fifth, the book offers numerous insights and examples from the authors' own academic research and high-tech industry experience.

The Audience

The book is written for second-year graduate and advanced undergraduate business students in all functional areas as a required course—Strategies for High-Tech Firms—or as an elective. Marketing and management majors will find it particularly useful.

Students who use this book in a course should have completed principles courses in economics and functional areas of business. No other prerequisites are required. The book is also targeted to the general business audience interested in high-tech markets as well as students in executive MBA programs and short-duration seminars.

ORGANIZATION OF THE BOOK AND OVERVIEW

The book is organized into eight chapters, each with an overview, key terms, and discussion questions.

Chapter 1 serves as a foundation for the rest of the book. It discusses unique economic problems faced by high-tech firms in appropriating the fruits of their innovative efforts; various forms of monetizing intellectual property; key drivers of strategy in high-tech firms; and determinants of appropriability regimes and their implications for marketing strategies. In addition, drawing upon the evidence from the pharmaceutical and software sectors, the chapter suggests potential application of the concepts and frameworks discussed.

In chapter 2 we discuss the traditional tools for analyzing strategic marketing issues and how they can be adapted to the high-tech environment. Specifically, we discuss Porter's Five Forces model, Ansoff's growth strategy matrix, the Boston Consulting Group's (BCG) portfolio matrix, the experience curve, and the product life cycle. One or more of these analytical tools are referred to throughout the book as appropriate.

Chapter 3 discusses how pricing strategies for high-tech products and services are guided by several features that distinguish them from other products and services. The chapter then provides a detailed discussion of examples of pricing strategies from three high-tech sectors—pharmaceuticals, commercial jet aircraft, and software. In addition, the chapter introduces and discusses the concept of disguised price discrimination, bundling, experience curve pricing, penetration pricing, bandwagon effects, and pricing of technology.

Chapter 4 reviews the main types of legal protection for high-tech innovations under intellectual property laws, including copyrights, patents, trademarks, and trade secrets; and explains the relative strengths and weaknesses of these protections as part of high-tech firms' strategies to reap as much of the economic value from their innovation as possible.

Chapter 5 discusses inter-firm collaborations, which have become an increasingly important strategy for high-tech industries because complex technologies are often too cumbersome, too interconnected with other technologies, and too expensive for any single firm to master alone. The resulting trend toward cross-licensing, strategic technological alliances, and joint ventures, discussed in the chapter on collaborations, may be viewed as an indicator of the emerging open innovation (OI) paradigm. The chapter reviews the ways in which various forms of inter-firm collaboration help the collaborating firms to obtain access to externally developed technologies that complement their internal technological and marketing resources.

In chapter 6 we consider product strategy issues that are particularly important to high-tech firms. They include: co-alignment of technology-based product concept with segmentation, design and standards, diffusion of innovations, branding innovations, marketing of technologies, open innovation, and platform strategy.

Chapter 7 discusses how the use of technology by firms and entire industries in their businesses can produce social costs and social benefits—known as negative and positive externalities—and how high-tech firms are dealing with these externalities. We use two case studies—the impact of high-tech firms' behavior on the environment, and individual privacy—to demonstrate the application of the externality concept to ethical issues, brand name, and reputation of the firm.

Chapter 8 focuses on the rise of the emerging-country pharmaceutical industry in the context of the Agreement on Trade-Related Aspects of Intellectual Property Protection (TRIPs), the changing structure of pharmaceutical innovation, and concern over the high cost of prescription drugs. It addresses two related issues rather broadly: (1) the strategic response of emerging-country pharmaceutical companies—with focus on India—to the new patent regime that recognizes and enforces product patents; and (2) its implications for the developed world's multinational enterprise (MNE) strategies—with respect to collaborations, outsourcing, and pricing, for example—vis-à-vis the developing-country markets and the ultimate threat to their own home markets. Thus the chapter presents an integrated view of marketing, economic, and legal issues concerning intellectual property rights in the context of the global pharmaceutical industry, referred to as the biopharmaceutical industry. The chapter considers these issues against the backdrop of continuing public policy debates about health.

ACKNOWLEDGMENTS

This book could not have been written without the help, support, and encouragement of students, colleagues, reviewers, friends, and the publisher. First, we fondly acknowledge the contribution of our friend and colleague Professor Eric Borg of Soderton University for many useful discussions in the early development of the book. We thank him especially for the suggestion that ethical and environmental issues should be part of any book on high-tech firms. We would like to thank Hayley Stewart and Pranav Patel, both former students at LIU Post, for their excellent assistance in research as well as the production end of this book. Aditya Deshpande assisted in the production of the manuscript in the final stage with great care. The book would not have been possible without their dedication to the project. We owe a debt of gratitude to Professors C. Jayachandran and Ram Subramanian of Montclair University for their incisive comments from which we benefited greatly. We thank our friends and former colleagues Jamie Howell and Jonathan Goldberg and Professor Jongtae Shin at LIU Post who reviewed major portions of the manuscript and made several helpful suggestions; and Professor Hyun Shin for helpful comments. We also thank the editorial team at M.E. Sharpe, Inc.—Patricia Kolb, George Lobell, and Kim Giambattisto—for guiding us through the early stages of preparing the manuscript for production; Stacey Victor for managing the production of the book; and Susanna Sharpe for excellent copyediting.

This book, as other similar works, stands on the shoulders of many great scholars whose works we cite. Ultimately, it is to them that we owe our greatest debt of gratitude. They include, in alphabetical order: David A. Aaker, Kenneth J. Arrow, George S. Day, William D. Nordhaus, Michael E. Porter, Everett M. Rogers, David J. Teece, and Oliver E. Williamson.

Finally, P.M. Rao thanks Jean Craig for her unfailing support and encouragement and most of all for her affection and friendship, which made a difference.

Strategies for High-Tech Firms

1 Sustaining Competitive Advantage in a High-Tech Environment

A Strategic Marketing Perspective

If inventions are completely inappropriable, no profit-maximizing competitor will produce an invention because increases in productivity would be instantaneously erased by a fall in price, and the firm would suffer losses to the extent of its research outlay.

– W. D. Nordhaus, 1969, p. 39

Chapter Overview

This chapter examines how high-technology firms can create sustainable competitive advantages despite the problems they face in appropriating the benefits of their innovative effort. Drawing upon the evidence from the pharmaceutical and software sectors, what follows is an overview of several relevant topics, including key drivers of strategies in high-tech markets and their implications for strategic marketing decisions.

Conceptual Context

Fundamental Problems Faced by High-Tech Firms

The fortunes of high-tech firms depend on investments in intangible capital, which includes intellectual capital as well as marketing capital. While the importance of intellectual capital—R&D capability, human capital, and the like—has been well established, the central role of complementary marketing capital in the process of innovation, in the form of brand name and other marketing assets, requires greater understanding. The inability of high-tech firms to increase the appropriability of the fruits of their innovative efforts, because of imitation, is a particularly difficult impediment to turning their technological competencies into sustainable competitive advantages.

The shift toward investing more in intangible capital has increased the importance of marketing strategies because intellectual capital is expensive to produce but cheap to copy. This is the classic problem of appropriability noted, for example, by Arrow (1962), Nordhaus (1969), Willimson (1975), and Teece (1987).

Unfortunately, legal protection of intellectual capital by means of patents, copyrights, and trade secrets does not work well in practice. "Inventing around" patents is a common practice that often can be accomplished at relatively modest cost. Pioneering studies by Mansfield (1985) and other scholars suggest that strategic information about product development usually lands in the hands of competitors within 12 to 18 months of introduction, and 60 percent of patented innovations are imitated within four years. Classic examples of innovators losing to imitators include the EMI CAT scanner, the Bowmar pocket calculator, and the Xerox office computer (Teece 2000). Furthermore, as shown in Figure 1.1, evidence produced by Agarwal and Gort (2001)—admittedly from a nonrandom sample of forty-six major product innovations—suggests that imitation by competitors is not only rapid but also has been growing consistently for more than a century.

Figure 1.1 **Interval Between the Introduction of an Innovation and Competitive Entry, 1887–1986**

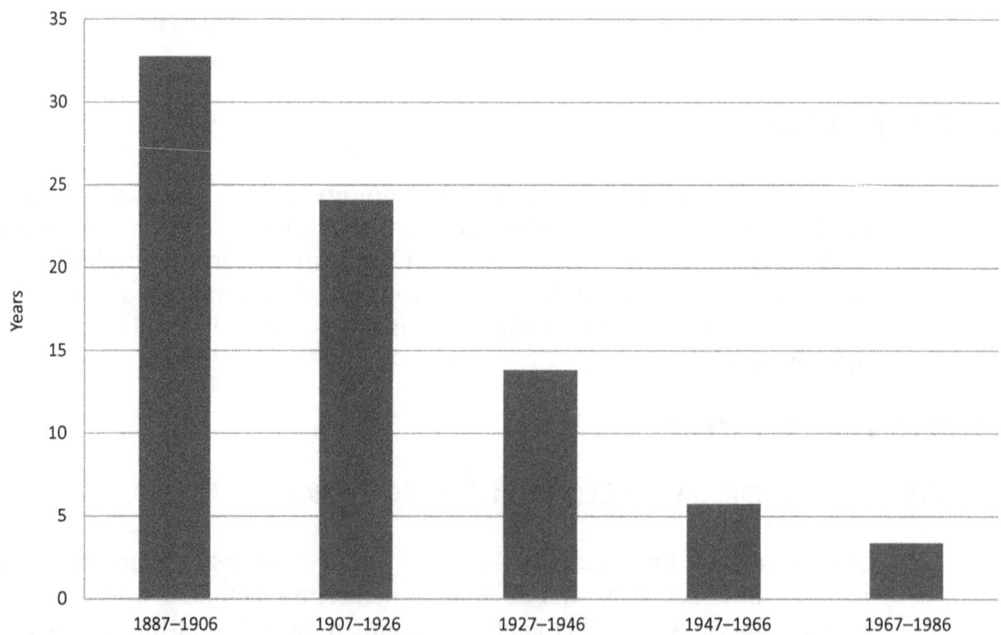

Source: Agarwal and Gort (2001).

DEFINITION OF TERMS

While it is not the purpose of this chapter to resolve the many definitional issues in the business and economic literature concerning technology, some key terms need to be defined at the outset. Although the term *intellectual property* (IP) can be defined to include a wide range of intangible assets, for purposes of this book it refers to all technology-based intangible assets of a firm—an idea or a design for a new product or a process, a computer software package, and the like—that may be protected as property under the legal framework that includes patents, copyrights, trademarks, and trade secrets (Rao and Klein 1994).

The term *high-technology firm* refers generally to a company with a relatively high level of research and development (R&D) intensity (the ratio of R&D expenditures of a firm to its sales) or simply a high level of R&D. One of the well-known shortcomings of this definition is that R&D expenditures represent inputs to the innovation process of a firm, not the outputs of that process. Still, R&D is one of the widely used indices of innovative activity of a firm in the business and economics literature (Griliches 1984). Moreover, it is particularly appropriate for this chapter because increasing the returns on R&D through marketing strategies is the central issue addressed here.

The term *high-technology products and processes* refer to products and processes that require, for their production, significant investments in R&D. The U.S. Department of Justice makes a distinction between *markets for technologies* and *markets for innovations* (U.S. Department of Justice 1995). *Markets for technologies* refer to existing markets for licensed intellectual property. *Markets for innovation*, on the other hand, are markets for future technologies, which include contract research, R&D joint ventures, technology alliances, and technical services.

MANY FORMS OF MONETIZATION

In modern economies, monetization of IP by firms takes several forms (see Rao 2012). These are now discussed.

PHYSICAL PRODUCTS

Perhaps the most common form of monetization involves incorporating the technology into a physical product—sometimes referred to as embodied technology—and marketing it. This is typical of innovating firms producing and marketing goods. In short, monetization of IP occurs in the product markets through sale, lease, and rental of products in which the technologies are embodied. That the embodied technology remains the dominant means of monetization can be seen from the fact that innovative effort in the form of R&D input is still largely concentrated in manufacturing. In 2009 in the United States, about 76 percent of company-funded R&D by U.S. multinational enterprises (MNEs) was in manufacturing (Barefoot and Mataloni 2011). As might be expected, judging from data on top U.S. patent holders, a similar concentration of R&D output in the form of patents appears to be present in the manufacturing sector (U.S. Patent and Trademark Office 2005). This is despite the fact that manufacturing in the developed world's economies accounts for an increasingly small fraction of output and employment, and the nonmanufacturing sector (outside of telecommunications and electric utilities to a limited extent), which was once devoid of much innovation, is now a rapidly growing center of innovation mainly in the form of software development. Note that monetization of IP—whether it is embodied into products or not—requires significant investments in complementary marketing assets in the form of building and sustaining a difficult-to-imitate brand name and reputation of the firm, establishing relationships with customers and distributors alike, and various promotional efforts in order to maximize the return on R&D investments.

PRODUCTION PROCESSES

A second form of monetization of IP occurs when an innovating firm employs its new technology in the production process, which reduces the cost and/or improves the quality of the end products or services it produces and markets. Note that process innovations—especially industrial processes—are generally not patentable. However, processes are difficult to replicate. Legal protection of process innovations mainly takes the form of trade secrets.

LICENSING

A third form of monetization of IP is through the licensing of technologies. This has become a significant and growing means of monetizing IP. Cross-border technology licensing has grown dramatically in the recent period. Between 1999 and 2009, U.S. receipts from licensing of industrial processes and computer software to other countries and payments to other countries have grown at an average annual rate of about 25 percent (see Table 1.1).

Moreover, as shown in Table 1.2, licensing revenues account for almost 100 percent of revenues for a technology firm like Rambus, which specializes in an innovative computer memory design called RDRAM; a vast portion of the revenue stream of software firms like Microsoft (86 percent); and a significant portion of a diversified firm like IBM (23 percent). Note that there are no physical products involved here. Technology is disembodied from the product. A variant of technology licensing is cross-licensing (discussed further in chapter 4), which involves companies exchanging technologies through the licensing process and netting out the difference in terms of valuation. Teece (2000) describes in detail various methods of valuation in the technology licensing process.

SALE OF TECHNOLOGIES

A fourth form of IP monetization occurs when technologies are sold outright. This form of monetization is less common than licensing. A recent dramatic example involved acquisition of Motorola Mobility, a subsidiary of Motorola, by Google for $12.5 billion (Richtel and Wortham 2011). The assets acquired by Google were simply the patents held by Motorola Mobility. Thus, Motorola, the parent, monetized

Table 1.1

U.S. Receipts and Payments of Royalties and License Fees, 1994 and 2009 ($ in millions)

	1999	2009	CAGR
Total, All Sources	$50,044	$115,021	8.67%
Receipts	39,670	89,791	8.59
Payments	10,374	25,230	9.49
Industrial Processes and Computer Software	9,687	93,128	25.27
Receipts	7,602	71,660	24.04
Payments	2,085	21,468	26.12

Source: Bureau of Economic Analysis (2011).

Table 1.2

Licensing Revenues of Selected High-Tech Firms, 2010

Firm	Revenue Sources	Revenue ($ in millions)	Percentage of Total Revenue
Rambus	Royalties from patent licenses	288.4	89.18%
	Royalties from technology licenses	31.8	9.83
	Total licensing revenue	320.2	99.01%
	Total revenue	323.4	
Microsoft	Windows Operating Systems	19,494	31.20%
	Microsoft Office and other software	19,076	30.53
	Servers	15,378	24.61
	Total licensing revenue	53,948	86.34%
	Total revenue	62,484	
Qualcomm	Total licensing revenue	4,011	36.52%
	Total revenue	10,982	
IBM	Software licensing revenue	22,485	22.68%
	Total revenue	99,120	

Source: Hoover's Online Database, 10-K Annual Reports.

IP held by its subsidiary, Motorola Mobility, in the form of a patent portfolio, a rich trove of 17,000 patents.

Market for Technologies

Note that licensing, cross-licensing, and sale of technologies suggest the presence of a "market" for technologies. Indeed, Baumol (2002) and Arora, Fosfuri, and Gambardella (2001), among others, argued about the emergence of a growing market for technologies. An interesting twist on this occurs when the acquiring company's motive has more to do with the value of human capital in the form of hard-to-get talented people employed by the firm to be acquired than the latter's patent portfolio. This is especially the case when the firm to be acquired is somewhat of a start-up with its financial feet not firmly on the ground.

SECURITIZATION OF IP

A fifth form of monetization involves securitization of intellectual property. While IP-backed securitization remains a small fraction of total asset-backed securitizations, Hillery (2004) documents several deals involving patents, copyrights, and even trademarks. One such deal was the pharmaceutical patent royalty securitization of variable funding notes in 2003 by the Royal Pharma Finance Trust (a $225 million deal involving a pool of patents on 13 drugs from 12 companies and Memorial Sloan-Kettering Cancer Center). Note that the deal can be structured so that the owner of IP continues to have an equity component of the royalty stream, while the buyer participates on the debt side at rates far lower than the typical venture capital lender. One way to view securitization is that it makes the "market" for IP more efficient than it would be otherwise.

ACQUISITION OF PATENTS

A sixth form of monetization involves firms that acquire patents from failed companies and independent inventors—often cheaply—and then "monetize" them by bringing infringement law suits against companies whose products are related to such patents. Note that such firms are not in the business of creating any technologies or making any products themselves. They specialize in recovering damages and collecting royalties on unauthorized use of their patent rights. The suit brought by NTP, a small, Virginia-based company, against Research in Motion (RIM) for violation of its wireless e-mail related patents, which claimed over a billion dollars in damages, is a case in point. NTP is a patent-holding company that does not make or sell products. Ultimately, RIM settled the case with NTP for $612.5 million (Somaya, Teece, and Wakeman 2011; Rao, Klein, and Chandra 2011).

Patent Trolls

Such firms are inelegantly called patent trolls, meaning that they are in the business of gaming the patent system by taking advantage of its inherent weaknesses. However, it is important to note that patent trolls do create markets for inactive and/or marginal patents and provide some reward to the original owners, which might not exist in their absence.

CLASSIC SOURCES OF MARKET FAILURE

High-tech firms present special problems for marketing strategy because of certain unique features associated with technological innovation. The three classic characteristics of technological innovation observed by Arrow (1962) are appropriability, uncertainty, and indivisibility; these are sometimes referred to as sources of "market failure" and are independent of the organizational environment in which it takes place.

APPROPRIABILITY

The fundamental marketing-strategy issue for a high-tech firm is how to minimize the appropriability problem. Only when a firm is able to minimize this problem can it turn its technological competencies into a sustainable competitive advantage (SCA). Despite the growing recognition of this issue by many industrial organization and marketing scholars (e.g., Ghemawat 1986; Teece 1987, 2000; Capon and Glazer 1987; Glazer 1991; Aaker 1998; John, Weiss, and Dutta 1999; Granstrand 1999; Vinod and Rao 2000; Mohr 2001), it has yet to take a central place in the strategic marketing literature.

Creating and sustaining a brand name and other difficult-to-imitate marketing strategies is the chief means by which successful innovators protect themselves against imitation. In 2008, economy-wide expenditures on advertising—which, according to some estimates, are less than a third of total expenditures on promotion—comprised *two thirds* of the industry-funded R&D at $267 billion or 1.9 percent of gross domestic product (GDP), versus $403 billion or 2.8 percent of GDP, respectively (based on data from Economic Report of the President 2012; see also Kotler and Keller 2008; National Science Foundation 2012; and U.S. Census Bureau 2012).

All forms of IP monetization suffer, albeit to varying degrees, from the appropriability problem due to reverse engineering and imitation around patents, competitors hiring away employees with trade secrets, and unpaid-for leakages to the licensees and any third parties whom the licensor did not intend to have access. On the other hand, rapid growth of technology licensing, monetization through sale of patent portfolios (e.g., Motorola Mobility), and securitization are indicators of the emerging, albeit highly imperfect, market for technologies.

As noted earlier, appropriability arises because innovation, which is often costly to produce, is imitated cheaply. As a result, firms operating in competitive market structures have difficulty realizing the full benefits of their innovative efforts and thus maximizing returns on R&D investments. This is part of the general problem of externalities or spillovers (not to be confused with network externalities) associated with knowledge assets. The appropriability problem poses the greatest challenge and the greatest opportunity for marketing strategy. It affects virtually all aspects of a firm's marketing strategy, including global strategy.

Social and Private Rates of Return

Competitive imitation is an important source of the appropriability problem and refers to the difficulty firms face in their ability to appropriate or benefit from the fruits of their innovative effort.

Innovation through investments in R&D often produces return to society—the social rate of return (SRR)—far greater than the return to innovator—the private rate of return (PRR)—because, as noted earlier, firms are unable to internalize or capture the full benefits of the fruits of their innovative effort. The extent of divergence between SRR and PRR is evidence of the extent of the appropriability problem, a source of market failure in the production of innovation, which leads to underinvestment in R&D. This, too, has been well documented in the economics literature. According to

Jones and Williams (1997), the social rate of return on R&D—estimated by various scholars—is about 30 percent, which is considered a lower bound, and about three times higher than the private rate of return. For a detailed discussion of estimated annual rates of return to R&D according to numerous studies, see Nadiri (1993). The greater the divergence between SRR and PRR, the less the extent to which a firm is able to appropriate the fruits of its innovative effort.

Schumpeterian Profits. When innovative firms are able to achieve PRR above the risk-adjusted rates of return on their investments in R&D, they are said to be enjoying super-normal profits, which Nordhaus (2004) refers to as Schumpeterian profits. Put differently, realization of super-normal profits means firms are able to capture a piece of the SRR as well. In his pioneering empirical study of the extent of Schumpeterian profits in the U.S. economy over the 1948–2001 period, Nordhaus provides a macro estimate and concludes that "only a miniscule fraction of the social returns from technological advances over the 1948–2001 period (*little over 2 percent*) was captured by producers, indicating that most of the benefits of technological change are passed on to consumers rather than captured by producers" (italics added, p. 1). Schumpeterian profits were only 2.8 percent of U.S. corporate profits during the same period. To be sure, appropriability varies greatly across industries and firms within the same industry and over time. Pharmaceuticals enjoy a high rate of profit, at least during the patent life, and farming and farm-equipment sectors have a low rate, despite their rapid productivity growth. Apple apparently has been enjoying extra-normal profits, at least for the time being, on its product line of iTunes, iPods, iPhones, and iPads.

UNCERTAINTY

Although uncertainty is a characteristic of all economic activity, it is so severe in the early stages of innovative effort that it is impossible to write enforceable contracts. As a result, the functioning of the markets for technologies and innovations is severely impaired and the problem of *tradability* arises. However, greater and better integration of marketing with R&D can reduce what Koopmans (1957) calls secondary uncertainty, which arises from lack of communication between decision makers about concurrent decisions and plans (primary uncertainty has to do with the state-contingent kind of uncertainty such as random acts of nature). Moreover, strategic and timely information about customers, competitors, and suppliers can improve the prediction of outcomes. John, Weiss, and Dutta (1999) suggest that greater uncertainty about the evolution of technology favors short-term licensing over outright sales of technology. This is because estimating the net present value of the former is easier. They further suggest that such an uncertainty favors modular designs over optimized designs, because the former allow advances in one part of the system to be incorporated without abandoning the whole system.

INDIVISIBILITY

Indivisibility arises when economies of scale (due to lumpiness of R&D projects) and economies of scope or economies of joint production and marketing (due to technical and marketing interconnectedness between various subsystems) are present in the innovation process. Lumpiness of a project means it is difficult to break it up into smaller parts. As a result, high-tech firms are often required to make larger upfront R&D investments (i.e., high fixed costs leading to scale economies) in large projects than they would otherwise.

While both scale and scope economies present problems for normal market transactions, technical interconnectedness requires close and continuous communication and coordination between R&D, manufacturing, and marketing (Teece 2000). Lumpiness, on the other hand, could result in the average size of the R&D projects being too large for many firms to undertake without joint ventures or alliances.

Teece (2000) notes four more characteristics of technological innovation as representing an emerging consensus among scholars: *path dependency, tacitness, cumulative nature,* and *irreversibilities.*

TACITNESS

Tacitness has to do with the fact that some knowledge created in the innovation process is not articulable and codifiable. Such knowledge is generally firm-specific and resides in the organization's ways of coordinating and managing tasks and cannot be seen in blueprints or manuals (Winter 1987). This makes transfer of technology to external organizations and tradability difficult, but at the same time makes imitation costly. John, Weiss, and Dutta (1999) posit that greater tacitness favors optimized designs over modular designs—the latter assumes easy transferability of know-how—and that it favors the sale of systems and subsystems over the sale or the license of know-how.

PATH DEPENDENCY

Path dependency refers to the idea that a firm's new product and process developments are likely to be determined by the technological competencies of its past. This limits the firm's ability to venture into new trajectories of research without resorting to acquisitions, contract R&D, alliances, or other forms of external technology partnering.

CUMULATIVE NATURE

Closely related to the concept of path dependency is the notion of cumulative nature of technological innovation. The idea here is that a firm cannot go too far in creating technological innovation without the cumulative knowledge base—much of it tacit and firm specific—it builds over time. While tacitness and firm specificity of technology protect it from easy imitation, they also create difficulties for market transactions involving transfer of technology.

IRREVERSIBILITIES

Irreversibilities in technological innovation occur when R&D expenditures are specific to the firm and have no alternative uses. They are sunk costs. Therefore, expenditures on R&D, especially money spent in the early stages of the innovation process, are best viewed as providing valuable information and creating options for further investments in production and marketing (Dixit and Pindyck 1995).

NETWORK EXTERNALITIES

Some authors (John, Weiss, and Dutta 1999; Rohlfs 2001) include *network externalities*—sometimes referred to as *external demand-side increasing returns* or *external demand-side scale economies*, in contrast to internal supply-side scale economies—as a special characteristic of technological innovation. Network externalities are also known as *bandwagon effects*. Network externalities arise when the value of a product or service to an individual user rises as the number of users increases (e.g., telephone networks, compatible computer networks that can share data and software). They follow Metcalf's law, which states that the value of a network increases as the square of the number of users. Teece (2000) refers to network externalities as direct network effects. In order to succeed, bandwagon products and services need to reach a critical mass, without which a *start-up problem* arises. Indirect effects occur when the value of network products is increased by the creation of large complementary markets (the larger the installed base of computers, the greater the volume and variety of software produced by third parties) and by the availability of large numbers of people trained to service it (the larger the installed base, the lower the cost of service). Rohlfs (2001) refers to these indirect network effects as complementary bandwagon effects.

Network products have important implications for marketing decisions concerning pricing, industry standards, and more. Another feature of technology markets is *unit-one costs*, which simply refers to the very high cost of producing the first unit and the near-zero cost of reproducing additional units, such as is the case with much applications software and movies on DVDs. A stylized summary of the key drivers of strategies in high-tech markets and their implications for strategic marketing decisions is shown in Table 1.3.

DETERMINANTS OF APPROPRIABILITY

Of all the characteristics of technological innovation discussed above, appropriability is the only one that affects virtually all aspects of a firm's marketing strategy. This section examines the key determinants of appropriability, drawing roughly and broadly upon the conceptual frameworks developed by Teece (1987, 2000), Teece and Pisano (1994), and Winter (1987). The importance of the appropriability problem for innovator firms cannot be overstated. David Teece (1987, p. 185) put it best:

> The message is particularly pertinent to those science and engineering-driven companies that labor under the mistaken illusion that developing new products that meet customer needs will ensure success. It may possibly do so for the product but not for the innovator.

Modern technological innovation is R&D intensive. While economy-wide company-funded R&D, as noted earlier, accounts for about 2.8 percent of GDP, the ratio of R&D to sales of companies in some high-tech sectors such as pharmaceuticals is

Table 1.3

Key Drivers of Strategies in High-Tech Markets: Implications for Strategic Marketing Decisions

Key Drivers	Implications
Appropriability Regime	
IPRs: Depends on strengths of patents, copyrights, etc.	Imitibility: Depends on the nature of technology product vs. process, know-how, tactic vs. codified
Entry decisions: Where, how, and when to compete?	Pricing and branding decisions
Tradability	Vertical positioning decisions: sell product or license technologies
Externalities	
External demand-side scale economies Bandwagon effect: Metcalf's law	Product design: Platform vs. targeted; modular vs. optimized
Network compatibility: standards, lock-in problem	Product offering: Bundled vs. unbundled
Unit-one costs	Pricing, product, and positioning decisions
Experience curve effects	Product and pricing decisions
Complementary marketing assets	Customer relationships, marketing, supply chain relationships, service, promotion, branding, and promotion decisions over the PLC
Expectations and uncertainty over future improvements in technology	Migration decisions
Start-up problems	Need to achieve critical mass of extraordinarily good products and services, and stand-alone applications (e.g., time-shifting applications of TV programs made possible by VCRs) would mitigate this problem

six times the U.S. average (National Science Foundation 2012; and Pharmaceutical Research and Manufacturers of America [hereafter PhRMA] 2010).

The process of innovation entails both technical and commercial risks. Empirical evidence suggests that commercial risks are far greater than technical risks. Although most R&D projects reach completion, only about a quarter of them ever achieve commercial success (Mansfield et al. 1997). With the firm's large investments in R&D—and sometimes its very existence—at risk, such as in the biotechnology field, how does a firm deal with the problem of imitation by competitors? Typically, a firm uses one or more of the following means: employing legal protection of IP, decreasing the replicability of the innovation, and boosting the complementary marketing assets.

LEGAL PROTECTION OF INTELLECTUAL PROPERTY

While chapter 4 is devoted to a lengthy discussion of IP from a legal perspective, in this section we provide a brief overview of the economic rationale underlying the legal protection of IP and the relative effectiveness of various instruments.

In the United States, the legal framework for protecting intellectual property consists of four major instruments: *patents*, *copyrights*, *trademarks*, and *trade secrets* (Bensen and Raskind 1991). Both patents and copyrights, because of their foundation in the U.S. Constitution and in various federal statues, constitute the highest form of legal protection. Semiconductor protection, derived from the Semiconductor Chip Act of 1984, is closer to the copyright regime.

A product technology is protected by patents (for a period of twenty years) and nonpatentable production/process technology is protected by trade secrets. Software is now protected by copyrights as well as patents. Software patenting is a relatively new phenomenon and has grown rapidly in the last fifteen years or so. Company names, logos, slogans, symbols, and brand names are covered by trademarks and copyrights.

Different instruments of legal protection can substitute for each other (e.g., patents and trade secrets) as well as complement or reinforce one another, as is more often the case. As Granstrand (1999) suggests, it is useful to look at different business elements or components that comprise a business system in a product area in relation to the legal instrument(s) applicable to each element or component. Thus, the development of a new generation of mainframe computers would most likely involve virtually all of the business elements (i.e., product technology, process technology, software, and brand name) and hence all legal instruments. Even when the issue is legal protection for a single business element such as software, firms use all available legal instruments—patents, copyrights, and trade secrets. However, patents remain the dominant legal instrument by which technological innovation is protected.

The traditional argument for legal protection of innovation through patents has to do with providing incentives to innovate by increasing the appropriability of the fruits of R&D efforts. Put differently, patents are intended to provide a temporary monopoly to the innovator. However, firms have come to use patents to gain competitive advantage in a variety of ways. These include: licensing in geographic markets or applications where the company is not present; cross-licensing; identifying and attracting partners for technology alliances; and various patenting strategies (e.g., "blanketing" or "flooding") intended to limit competitive entry into a given field of technology (Granstrand 1999).

Survey Findings on the Effectiveness of Patents

Despite some advantages that patents confer on innovating firms, survey findings by Cohen, Nelson, and Walsh (2000) shown in Table 1.4 indicate that in almost all industries, respondents report lead time and secrecy, and not patent, as most effective for a higher percentage of tacit innovations. Complementary sales service ranks higher than patents in six out of ten industries surveyed in the United States.

Similar surveys conducted by Grandstrand (1999) for Sweden and Harabi (1995) for Switzerland concluded that managements consider creating superior marketing and creating market lead times to be more effective than patents. Although Japanese firms place high value on "taking out patents to deter imitators (or to collect royalties)," when asked about various limitations of patents for technology protection, the

Table 1.4

Effectiveness of Appropriability Mechanism for Product Innovations

Selected Industries	Sample Size	Secrecy	Patents	Other Legal	Lead Time	Complementary Sales Services	Complementary Manufacturing
Food	89	59	18	21	53	40	51
Petroleum	15	62	33	6	49	40	36
Basic chemicals	35	48	39	12	38	46	45
Drugs	49	54	50	21	50	33	49
Machine tools	10	62	36	9	61	43	35
Computers	25	44	41	27	61	40	38
Electrical equipment	22	39	35	15	33	32	32
Semiconductors	18	60	27	23	53	42	48
Medical equipment	67	51	55	29	58	52	49
Auto parts	30	51	44	16	64	45	53
ALL	1,118	51	35	21	53	43	46

Source: Cohen, Nelson, and Walsh (2000), Table 1.

Note: Numbers represent mean response in terms of percentage of product innovations for each industry for which the type of protection shown in the columns is effective.

same firms placed the statement "competitors can legally circumvent patent rights or invent ways around them" at the top of the list (Granstrand 1999, p. 190). Thus, even the Japanese firms appear to view patenting, at best, as a defensive move against potential imitators. When managements of these companies were asked about the effectiveness of various means of ensuring appropriability of the results of product and process innovations, patents were considered to be least effective. Superior sales and service efforts received the highest scores, followed by lead time. And for process innovations, lead time, followed by superior sales, received the highest scores. Note that lead time is related to sales and service efforts as it helps build a whole new range of marketing assets, including superior sales and service infrastructure, well before competitors enter the market. Indeed, lead time gained in creating marketing infrastructure may well discourage competitive entry into the market.

The Case of Software

Among the high-tech sectors, nowhere is appropriability a greater problem than in the software sector, which is the backbone of the new economy. As noted earlier, software is protected by patents as well as copyrights. The problem goes beyond software piracy (unlicensed software units as a fraction of total units installed), which is estimated at about 43 percent of worldwide sales, bearing in mind that such estimates generally carry a high margin of error (Business Software Alliance and International Data Corporation 2010).

The real source of the problem is changing technology and customer demands, which make appropriability difficult. The issues include development of industry standards (standards based upon operating systems that permit portability of specific customer applications from one computing environment to another and provide customers the flexibility to "mix and match" hardware and software from different vendors); customer demand for greater control of software in order to create, manipulate, and manage applications to their own specifications; and modularizations of software development that make customer control of applications and common interfaces among systems from multiple vendors easier. In short, these trends have the effect of making legal restrictions imposed by software vendors on customers' use of their products untenable and appropriability difficult (Rao and Klein 1994).

The changing scope of copyright protection is yet another trend accentuating the appropriability problem in the software sector. The European Union (EU) allows a person to freely reproduce or translate the form of another's program code for the specific purpose of facilitating interoperability with an independently developed system. In the United States, while reverse engineering is not exempt from copyright protection, the case law appears to be moving closer to the EU position. Moreover, copyright protection for user and system interfaces is ambiguous (Rao and Klein 1994).

REPLICABILITY OF INNOVATION

Appropriability of technological innovation depends on the strength of intellectual property rights (IPRs) as well as the degree to which innovation is inherently replicable by competitors. As Teece (2000, p. 18) puts it, "Imitation is simply replication performed by a competitor." It is easier to replicate codified knowledge embodied in an innovation than tacit knowledge. Winter (1987) examines the relationship in terms of technology being tacit versus articulable (or codifiable); not observable in use versus observable in use; complex versus simple; and an element of a system versus independent. Note that the right end of each of the above taxonomic dimensions—articulable, observable in use, simple, and independent—indicates that replicability is easier and, ceteris paribus, appropriability lower. Technological innovations embodied in packaged software fit the dimensions of low appropriability.

However, it is important to note that codified knowledge by itself does not necessarily diminish appropriability. For example, patented drugs in the United States and in other industrialized countries enjoy a high degree of appropriability, even though the knowledge embodied in the chemical compound is not only codified but known to everyone by the time the drug is approved. This is because IPRs are strong and patent enforceability with respect to drugs is high. It is the absence of these conditions that makes software less appropriable.

It should be noted that while IPRs and replicability jointly determine the degree of appropriability, as the above discussion suggests, both are largely exogenous variables and therefore the firm has limited control over them. This is notwithstanding the benefits that could be derived from an aggressive policy of securing and enforcing IPRs and implementing a variety of secrecy measures internally.

STRENGTHENING COMPLEMENTARY MARKETING ASSETS

Complementary marketing assets are central to increasing the appropriability of the fruits of investments in R&D. Such assets comprise a whole range of cumulative investments: a well-trained sales force; superior after-sales service; promotion to create and sustain a brand name; channel development; carefully developed relationships along the entire spectrum of the supply chain; superior information systems concerning markets, customers, competitors, and more. In short, they include all marketing activities that help create and sustain what Porter (1985) calls advantage of the first mover. The essence of marketing strategy is the creation of competitive advantages—the more the better—with investments in difficult-to-imitate marketing assets.

Tacit and Codified Assets

However, marketing assets, like many other assets, fall on a continuum that ranges from highly tacit (e.g., the process of creating a successful channel relationship or a brand name) to highly codified (e.g., a detailed after-sales service manual).

Winter's (1987) taxonomic dimensions concerning technology apply equally well for marketing assets. The greater the tacit component in a firm's marketing assets, the less observable they are in use, and the more complex they are, the smaller the chance of imitation and the greater the effectiveness of marketing strategy in increasing the appropriability of innovations.

More recently, evidence, albeit indirect, of the importance of R&D and marketing to high-tech firms and their complementary relationship came from an important study by Hulten (2010). Hulten used familiar macro-level growth accounting techniques to explain sources of Microsoft's average annual rate of growth in real output of 30 percent over the 1988 to 2006 period and found that R&D and sales and marketing together explained—roughly in equal proportion—43 percent of Microsoft's growth. Microsoft's employment data reported by Hulten are consistent with the importance of R&D and marketing. In 2006, 28 percent of Microsoft's employees were in product R&D; 30 percent in sales and marketing; 17 percent in product support and consulting; and 10 percent in general administration. Note that only 3 percent of Microsoft's 71,000 employees were in manufacturing and distribution.

Brand Name

To cite another example of the role of complementary marketing assets in mitigating the appropriability problem, few would argue that Apple's ability to produce Schumpeterian profits in recent years has as much to do with the strength of its brand name—often personified in its CEO, the late Steve Jobs, and supported by huge investments in its own uniquely designed retail stores amply populated by educated sales people ready and willing to serve—as with its brilliant design and software. To elaborate on this point, Apple offshored all manufacturing of iPhone 4 to low-cost, high-quality Asian countries and still captured nearly 60 percent of the total value as a reward for its investments in design innovation and marketing assets (Kraemer, Linden, and Dedrick 2011).

Creating Sustainable Competitive Advantage

Imitating a successful brand strategy combined with relationship marketing is a lot more difficult than imitating a product design. One need only consider the

$65 billion brand value of IBM in 2010 (57 percent of the value of the assets on the balance sheet), bearing in mind the generally high margin of error associated with such estimates (see discussion in chapter 6). Day (1997) has noted five conditions that make an asset a source of sustainable advantage: it is *valuable* in that it produces superior customer value; it is *durable*; there is *causal ambiguity* (i.e., competitors cannot figure out how the source of advantage works); competitors find it *difficult to duplicate* even if they understand the source of advantage; and an innovator firm will deter efforts at imitation with a threat of *retaliation*. Others have suggested that skills or distinctive competencies in the form of a focus on total customer satisfaction, a focus on continuous innovation combined with "an all-hands effort" to achieve them, would be virtually impossible to imitate (Peters 1984).

It should be noted that in contrast to IPRs and replicability, a firm's marketing assets are endogenously determined and therefore are largely within its control. Thus, once the innovation is under way, complementary marketing assets become an intervening variable in determining the appropriability of technological innovation. The creation and development of complementary marketing assets mitigates the problem of loose IPRs and easy replicability.

However, the real task lies in converting these assets into superior positions of competitive advantage. What Day and Wensley (1988) call drivers of differentiation are particularly important. According to Day and Wensley, many of these drivers, which are policy choices about features, performance, level of advertising expenditures, and the like, are mediating events that determine the leverage an investment in a skill or resource has on differentiation rather than sources of advantage. Other important drivers of differentiation include linkage within the value chain (e.g., coordination between sales and service to improve the speed of order handling); timing, which gives first-mover advantages; interrelationships with other business; and learning.

Few, if any, of these drivers of differentiation would be effective without the aid of information and communications technology, which itself can be thought of as a driver of differentiation as well as a source of competitive advantage. Porter and Miller (1998) provide numerous examples of how information and communications technology (ICT) has permeated the value chain and transformed the way value-creating activities are performed and the linkage among them; transformed the product by vastly increasing the information component relative to the physical component of the product (e.g., electronic control of the automobile); changed the competitive scope by increasing a firm's ability to coordinate its regional, national, and global activities (e.g., coordination of the Asian, European, and U.S. editions of the *Wall Street Journal* by Dow Jones and Company); spawned new businesses (e.g., Western Union Easy Link service); lowered cost (e.g., the software behind Canon's successful low-cost copier assembly process); and enhanced differentiation (e.g., myriad travel services offered by American Express).

However, it is important to note that ICT is a two-edged sword. While it is a driver of differentiation and a source of competitive advantage, it can mitigate the appropriability problem (through encryptions and "firewalls," for example) or accentuate it through free flow of information, a feature inherent in its design. It was even argued elsewhere that in this age of supply-chain management—in which information sharing with suppliers, customers, and other external partners is central to the innovation process—ICT is more likely to accentuate the appropriability problem than mitigate it (Rao 2001a).

APPROPRIABILITY REGIMES

Although Teece (2000) recognizes the importance of specialized complementary assets, including marketing assets, for commercialization of innovations, his focus is on predicting the outcomes of the interdependence between the innovator (and the imitators) and the owners of such assets. Teece views the appropriability regime as a function of IPRs and inherent replicability. Figure 1.2 adds complementary marketing assets as a third, albeit an intervening, variable in the determination of appropriability.

Note that Figure 1.2 provides a highly stylized partial mapping of appropriability regimes that would result from all possible combinations of the three determinants. Moreover, the discussion of Figure 1.2 below assumes that the innovator firm has the greatest degree of control over the creation and deployment of marketing assets, some control over IPRs, at least in terms of aggressive enforcement policy, and little or no control over replicability.

Figure 1.2 **Key Determinants of Appropriability Regime**

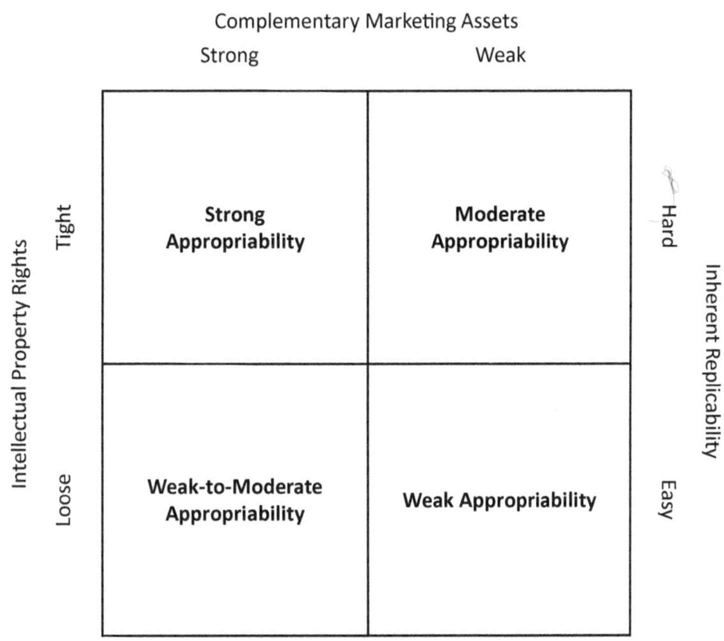

Source: Adapted from Teece (2000, p. 19). By permission of Oxford University Press.

STRONG APPROPRIABILITY REGIME

The upper left quadrant in Figure 1.2, with tight IPRs and hard replicability complemented by strong marketing assets, represents a strong appropriability regime. This is generally a winning combination for the firm. It keeps the potential imitators at bay and gives the firm a strong competitive advantage to start with. Moreover, a strong appropriability regime always offers the innovator firm an opportunity to increase the primary demand for the product through promotion. The burden of sustaining this initial advantage falls largely on successful implementation, over time, of the strategic marketing decisions associated with the creation and deployment of marketing assets. To be sure, strategy as well as implementation are product/market specific and need to be adapted to the changing technologies and customer needs. For example, a pharmaceutical firm specializing in patented prescription drugs would emphasize promotional strategy more than a software firm whose concern was product strategy in the context of rapidly changing technologies and customer demands.

Implications for Product Life Cycle

An innovative product in a strong appropriability regime, which the firm itself helps to create, could experience a textbook S-shaped product life cycle (PLC) under the usual assumptions. The firm has a temporary monopoly over its innovation by virtue of a strong patent or other forms of tight IPRs. Under these conditions, the innovator firm has a strong incentive to invest in complementary promotion as a barrier against competition. In the prescription drug segment of the pharmaceutical industry, for example, firms maintain a high level of promotional intensity (i.e., ratio of promotional expenditures to sales) during the patent period to build brand equity, the benefits of which are reaped during a post-patent period, when price competition from the generics is intense. Moreover, as suggested earlier, investment in promotion in the early stage of the PLC would have the effect of increasing the primary demand for the product.

A strong appropriability regime combined with heavy up-front investment in promotion could give the innovator firm greater flexibility in pricing. The expectation of a longer S-shaped PLC favors a strategy of skimming pricing, which can be viewed as price discrimination across different stages of the PLC. Note that price discrimination is generally more profitable than nondiscriminatory pricing.

With respect to product strategy, a strong appropriability regime favors radical innovation. The large fixed development costs or unit-one costs associated with such innovation favor platform design, which offers several variants of a core design to each market segment (John, Weiss, and Dutta 1999). The firm has a strong incentive to invest in long-term R&D, including basic research, and undertake large projects. The S-shaped PLC the firm faces is consistent with skimming pricing. The firm follows a distribution strategy that starts with selective distribution in the introductory stage of the PLC and moves gradually to intensive distribution in the later stages. A strong appropriability regime confers several other advantages on the innovative firm. The threat of new entrants is likely to be low, at least in the near term. Therefore, the firm has some latitude with respect to innovation time span and speed to market.

MODERATE, WEAK-TO-MODERATE, AND WEAK APPROPRIABILITY REGIMES

Moderate Regime

The upper right quadrant in Figure 1.2 presents the most interesting case. A regime of moderate appropriability is characterized by tight IPRs and hard replicability, complemented by weak marketing assets. It is within the control of an innovator firm to move to a strong appropriability regime (the upper left quadrant) by moving to a position of strong marketing assets by making additional investments in them. The return on this additional investment (e.g., promotional expenditures aimed at creating a brand value and/or channel development) is likely to be particularly large during the post-patent period when imitators enter the market. Not investing in such marketing assets increases the likelihood that the innovator will lose to imitators.

Weak-to-Moderate Regime

A combination of loose IPRs and easy replicability complemented by strong marketing assets produces a weak-to-moderate appropriability regime (lower left quadrant in Figure 1.2). Given the presence of strong marketing assets (e.g., a strong brand name and distribution system), aggressive litigation and secrecy strategies could move the innovator firm from a weak-to-moderate to a moderate-to-strong position. Thus marketing assets can mitigate the weaknesses present in the IPRs and replicability.

Weak Regime

A combination of loose IPRs and easy replicability complemented by weak marketing assets (lower right quadrant in Figure 1.2) produces a weak appropriability regime. Loose IPRs combined with easy replicability are a strong disincentive to invest in R&D, much less in complementary marketing assets. Aggressive and often costly litigation and secrecy strategies could overcome some of the problems associated with loose IPRs, but their effectiveness is uncertain. To the extent innovation takes place, the winning factor, in general, is the strength of innovators' marketing assets relative to those of imitators.

IMPLEMENTATION STRATEGIES

Regardless of the appropriability regime with which a firm is faced, the central role of complementary marketing assets in overcoming the problem of imitation is clear. However, the issue is devising an implementation strategy—bearing in mind that marketing assets constitute but one component of the firm's capabilities—that would create and sustain competitive advantages.

SPP FRAMEWORK

Day and Wensley (1988) and Day (1994 and 1997) have addressed this difficult issue from a marketing strategy perspective. First, competitive advantage needs to be examined in terms of its three components, namely, sources, positions, and performance outcomes—the "SPP framework." Note that this framework emphasizes integration of the capabilities approach (Teece 2000) and the strategic positioning view (Porter 1991) in a manner that the choice of capabilities to develop and investment commitments to make must be guided by the positional advantages being sought, given the competitive forces in the industry, the needs of the target segments, and the environmental trends (Day and Wensley 1988).

CAPABILITIES APPROACH

Second, Day (1994) suggests that the design of effective programs to create and sustain competitive advantage by delivering superior customer value comes from combining the capabilities approach to strategy with total quality management (TQM). Central to a successful implementation strategy are market-sensing capability and customer-linking capability.

Market sensing allows firms to identify the changing requirements of the customer early and respond with a continuous stream of innovative products and services. Customer linking has to do with interlinked skills, abilities, processes, and procedures needed to respond to customer needs quickly and with minimum overhead and internal transaction costs. Note that investments in ICT are critical to developing both market-sensing and customer-linking capabilities.

MULTI-PROTECTION SYSTEMS

Finally, the special problems of high-tech firms require that IP protection must become an integral part of corporate and marketing strategies. IP issues are too important to be left entirely to the legal departments. High-tech firms need multi-protection systems (patents, trademarks, and trade secrets) and total IP strategies, which include technology acquisition strategies (e.g., internal R&D versus external acquisitions), technology exploitation strategies (e.g., production and marketing of products versus technologies), commercialization strategies (e.g., patenting, superior marketing, and market lead times), and patenting strategies (Granstrand 1999). Also needed is a corporate-wide education of IP issues down to the level of accounting for IP, project evaluation, and pricing.

OPEN INNOVATION AND MONETIZATION OF INTELLECTUAL PROPERTY

Open innovation as a means of monetizing IP has its roots in the notion of user innovation proposed by von Hippel (2005), among others. The trend toward open innovation has been discussed at length by Rao (2008).

OPEN INNOVATION PARADIGM

Several recent trends—the growing importance of innovation networks, increasing trend toward alliances, and licensing coupled with private-public nonprofit linkages—may be viewed as indicators of the emerging open innovation (OI) paradigm, which has received much attention in recent years (Chesbrough 2003). In contrast to the traditional vertically integrated (VI) closed innovation model, a company following the OI model commercializes internal as well as external ideas by utilizing outside as well as in-house pathways to the market. In the OI model, large companies are seen as becoming more and more like "market coordinators of increasingly distributed and vertically disintegrated value chains" (Christensen 2006, p. 46). And as Chesbrough (2003) suggests, the boundary between a firm and the surrounding environment becomes more porous, enabling innovative activities to move easily between the two. As such, OI is the antithesis of the VI model; it challenges the core competency perspective of corporate innovation, and it emphasizes research and production via collaboration and networks. Firms adopting OI recognize the value of external IP as well as profits to be made from others' use of its IP through licensing, joint ventures, and the like. Among the pharmaceutical MNEs, Pfizer is such a firm. It uses external innovations in its product mix and markets internal as well as external innovations (West, Vanhaverbeke, and Chesbrough 2006).

Still, open innovation presents a challenge to monetization of IP in the case of both biotech firms and firms operating under the open source software model. The OI model goes against the traditional approach of maximizing profits through innovation from strong IPRs (Christensen 2006; West and Gallagher 2006). Moreover, Chesbrough (2011) argues that product-driven companies are finding it harder to differentiate because of rapid commoditization made possible, in part, due to numerous software products. For example, Motorola's highly successful RAZR cell phone fell by the wayside within three years due to pressure from Apple and Google, who are building platforms and services in collaboration with thousands of companies. The solution is to transform physical products into platforms with a variety of services that are created through open innovation. A different kind of transformation occurred at Amazon, which started in 1994 as an online bookseller and within a period of just over ten years created and sustained one of the most successful online retail industry platforms in the world and leveraged its technology to move into providing Web services to third-party retailers (Isckia and Lescop 2009).

OPEN SOURCE MOVEMENT

Open innovation is often compared, albeit loosely, with the "open source" movement led by IBM and other companies in software development. (Although OI and open source have some elements in common, it is important to note that the open source movement is unique to the software sector and is driven by network externalities and other factors that are not present in many sectors.) In the open source model, developers are free to access and build upon the efforts of others without paying royalties. Investment by firms in open source projects to gain competitive advantage through pooled R&D is consistent with open innovation.

Merck's Gene Index

Pisano (2006) provides an interesting story about the case of Merck's Gene Index in this context. It suggests that, contrary to the traditional view of IP strategy, a firm with strong downstream marketing assets may find the value of those assets is higher in a regime of weak IPRs rather than strong ones. This was the case of Merck's collaboration with Washington University to create a database of the human gene sequence and make all findings public within forty-eight hours. (This was in response to venture capital firms aggressively patenting genes associated with diseases in which Merck's research and marketing assets are strong.) Merck viewed its interests as best served by preventing privatization of genes that could result in lockout in research areas central to its survival. Thus, a firm with strong R&D and downstream marketing assets can gain more by adopting the collaborative network-driven OI model even if it means a significant amount of proprietary knowledge spills over into the public domain. The advocates of the OI model suggest that firms adopting the model may find that it is compatible with a high degree of appropriability even as it dilutes IPRs.

Open Source Software Movement

Schwarz and Takhteyev (2009) argue that the intrinsic inefficiencies of proprietary software led to alternative mechanisms—such as the open source software (OSS) movement—that provide software as a public good. One source of inefficiency is the holdup problem that occurs when users of software must make complementary investments to modify and/or customize the software for their unique needs, and when the success of such investments depends on the cooperation of software vendors. Another source of inefficiency has to do with production of software becoming more expensive when the source code is not open to the end users. The authors explain why open source dominates certain sectors of the software industry and not others. For example, the top ten programming languages all have an open source implementation because they require substantial complementary investments and users have the need and the capacity to modify the source code. On the other hand, none of the software for the top ten computer games is likely to need modifications, and is offered to users with few or no programming skills to modify it even when the source code is offered to them. Therefore, it is virtually all proprietary.

Open source operating systems and tools are being embraced not only by individuals, but they have become preferred systems and tools for Internet companies such as Amazon, Facebook, Google, and Yahoo! as well as IBM, Intel, and Oracle, among others. Corbet (2007) estimates that about one-third of the changes to a recent release of the Linux kernel were made by employees of only five companies (IBM, Novell, Intel, Oracle, and Red Hat). Schwartz and Takhteyev (2009) report that in 2001, IBM announced its plans to contribute $1 billion to the Linux open source operating system and later reported that the investment was recovered in one year. This is not surprising because, given the need for complementary investments and frequent modifications in the software, combined with their capacity to make modifications, these companies understand the individual and collective benefits in making software a public good. Therefore, monetization of investments in OSS occurs through products and services enabled by the software rather than software itself. The holdup problem is thus largely mitigated.

OPEN INNOVATION AND SOCIAL RATE OF RETURN

While open innovation—of which OSS is one example—has the property of increasing SRR, it simultaneously reduces appropriability and hence reduces the innovator firm's incentive to innovate. This is especially the case when an innovator firm lacks the depth and breadth of downstream products and services (like IBM, for example) that benefit from OI. Put differently, a firm like IBM has more to gain by playing the OI and the proprietary games at the same time, whereas a firm like Rambus specializes in the production and licensing of proprietary technologies for its existence and survival.

SUMMARY

Managers of high-tech firms must recognize the unique characteristics of technological innovation in formulating marketing strategy. This chapter has focused on the appropriability problem—a particularly important concern facing high-tech firms. It has suggested the importance of investing in complementary marketing assets to mitigate this problem. However, complementary marketing assets take different forms in different industries.

In the patented and branded segment of the pharmaceutical industry, which combines tight IPRs (at least in industrialized countries) with easy replicability, the emphasis is on heavy complementary promotion during the patent life of a drug. The complementary relationship between R&D and promotion in the pharmaceutical industry has been well established in an econometric study by Vinod and Rao (2000). Moreover, the data produced by the authors strongly suggest that a firm that recognizes the strategic importance of complementarity between R&D and promotion has a greater chance of survival than a firm that does not. The objective of such a strategy is to increase the primary demand for the product as well as build a brand name, which will help sustain the competitive advantage during the post-patent period when price competition from the producers of generic drugs is intense.

The software industry has different characteristics and requires a different approach. In general, the software industry is characterized by a low degree of appropriability due to inherent difficulties in separating innovative and protectable expressions of original ideas, such as a "user graphic interface" from unprotectable ideas such as an algorithm (Grimaldi and Torrisi 2001). Although software now has copyright as well as patent protection in the United States, because of the uncertainty surrounding the legal protection, the strategic emphasis is on the product itself.

Complementary marketing assets take the form of significant investments made by the firm to achieve the optimal level of codification in order to enhance appropriability. Note that, as with prescription drugs, codification, while it makes imitation easier, results in greater enforceability of copyrights and patents and therefore greater appropriability. Thus, packaged software, which is highly codified and growing in importance, has greater appropriability than custom software. Moreover, firms producing software products for PCs and the Internet have the opportunity to increase appropriability by exploiting network externalities through a pricing mechanism, which recognizes that the value of a product or service to an individual user rises with the number of users. Still, the emergence of open source software, which might be viewed as part of a broader trend toward the collaborative model of open innovation, challenges the conventional argument that the proprietary model of technology achieves the maximum level of appropriability.

THE MARKET FOR TECHNOLOGIES

The market for technologies (e.g., licensed IP, technology joint ventures, and alliances), which has been growing rapidly, presents even greater challenges and opportunities than the market for high-tech products. The extraordinarily rapid growth of licensing fees from industrial processes and computer software in recent years (25 percent per year) and the high proportion of revenues from licensing of technologies derived by a diversified high-tech firm such as IBM (about 23 percent), shown in Tables 1.1 and 1.2, support the view advanced by Baumol (2002) and Arora, Fosfuri, and Gambardella (2001) concerning the emergence of a market for technologies.

It is important to recognize the risks and opportunities present in the licensing of technologies. While licensing contributes to the current and future revenue stream of the firm, it also increases the risk—notwithstanding the legal protection written into contracts—of cannibalizing future profits due to the licensee competing in the product market. Sometimes this issue could go to the heart of a firm's strategy. Microsoft, with its PC-based software, generally follows the traditional approach to licensing, which gives the licensee access to executable code, but not the source code. By contrast, Linux vendors agree to license any line of source code—known as a "general public license"—which they produce against full access to the source code the licensee may develop. The Linux model gives each customer unrestricted access to the source code to make improvements and customize it to meet each customer's unique requirements (Grimaldi and Torrisi 2001). Microsoft and Linux vendors operate on different business models and cater to different segments of the market—Microsoft to the desktop operating systems' segment and Linux vendors to the server operating systems' segment.

The rapid growth of technology alliances since the early 1980s, which has been well documented (Duysters 1996; Rao 1999, 2001b), represents yet another indicator of the emerging market for technologies. Note that the notion of the market for technologies presents something of an anomaly for some celebrated theories of "market failure" in the production of technological innovation (Arrow 1962; Williamson 1975). Technology alliances, which represent a trend toward unbundling and outsourcing of corporate R&D, present risks and opportunities like any other form of technology trading. The gains from technology alliances are market access, technological complementarity, and reduction in innovation time span. The risks involve potential loss of control over critical technologies and, hence, low appropriability of innovations. Given the uncertainty and irreversibility associated with most R&D investments, technology alliances are best viewed as value-creating options. That is, alliances give a company the option but not the obligation to make additional investments in the future.

Innovative firms' global entry decisions are influenced by the relative strength of a country's IP rights. This is true notwithstanding the significant degree of harmonization of IPRs, which has been taking place under the World Trade Organization system. Ceteris paribus, countries with tight IPRs are attractive candidates for technology transfer through foreign direct investment (FDI)—including FDI in R&D, joint ventures, strategic alliances, and licensing. Countries with loose IPRs are poor candidates for such transfers. When countries have poor IPRs, the transfer of the newest and most effective technology through licensing is least favored (this is because of the fear that the know-how will fall into the hands of potential competitors), followed by FDI and joint ventures (Mansfield 1995). However, judging from evidence on the pharmaceutical industry (Rao 1998), a regime of tight IPRs alone would not do the job for the developed world's multinational firms. Global marketing strategies—for example, segment-specific pricing, creative branding strategies, and direct involvement with local partners—are likely to play a greater role in the decisions concerning technology transfer to countries with poor IPRs.

Finally, it is impossible to create and sustain competitive advantages without an implementation strategy. What Day and Wensley (1988) and Day (1994, 1997) suggested is still a good starting point. Central to a successful implementation strategy, especially marketing strategy, are market-sensing capability and customer-linking capability. Market sensing allows firms to identify the changing requirements of the customer early and respond with a continuous stream of innovative products and services. Customer linking has to do with interlinked skills, abilities, processes, and procedures needed to respond to customer needs quickly and with minimum overhead and internal transaction costs. Investments in ICT are critical to developing both market-sensing and customer-linking capabilities.

The fundamental issue in marketing strategy is how to create sustainable competitive advantage (SCA) for the firm. From a marketing strategy perspective, SCA rests on three rather obvious but important strategic imperatives. First, a firm's activities must meet users' needs. Second, a firm's product or service must be unique and the firm must leverage that uniqueness. Differentiation allows the firm to become a price setter rather than a price taker. Third, a strategy of differentiation must be difficult to replicate so that profits won't be competed away. The last point is particularly important for high-tech firms because, often, competitors can imitate easily and inexpensively. Still, increasingly, high-tech products and services define modern economies. Indeed, there is an emergence of the market for technologies themselves in the form of licensing, cross-licensing, alliances, and the like. All of them face the problem of spillovers that often benefit the competitors more than they benefit the innovator. The arguments presented in this chapter strongly suggest the importance of investing in difficult-to-imitate complementary marketing assets that will increase appropriability and returns on R&D investments. Replicating a successful brand strategy is lot more difficult than replicating a product design. The importance of marketing strategy in protecting intellectual capital cannot be overstated.

KEY TERMS

appropriability
demand-side scale economies
experience curve effects
externalities
imitability
indivisibility
intellectual property rights
irreversibilities
open innovation
open source software
patent trolls
path dependency
private rate of return
replicability
Schumpeterian profits
securitization of intellectual property
social rate of return
start-up problem
tacitness
tradability
unit-one costs

DISCUSSION QUESTIONS

1. What are the key drivers of strategies in high-tech markets and their implications for marketing strategy decisions?
2. Explain why technological innovation through investments in R&D is a source of sustainable competitive advantage as well as a risk for a high-tech firm.
3. Describe the basic characteristics of technological innovation and their implications for marketing strategy.
4. Explain what is meant by the appropriability problem faced by high-tech firms and discuss the key determinants of appropriability.
5. Discuss the central role of complementary marketing assets for high-tech firms.
6. What are the various forms of monetizing intellectual property?

REFERENCES

Aaker, David A. 1998. *Strategic market management.* New York: John Wiley & Sons.

Agarwal, Rajshree, and Michael Gort. 2001. First mover advantage and the speed of competitive entry, 1887–1986. *Journal of Law and Economics* 44 (April): 161–177.

Arora, Ashish, Andrea Fosfuri, and Alfonso Gambardella. 2001. *Markets for technology*. Cambridge, MA: MIT Press.

Arrow, Kenneth J. 1962. Economic welfare and allocation of resources for invention. In *The rate and direction of inventive activity*, ed. R. Nelson, pp. 609–624. Princeton, NJ: Princeton University Press.

Barefoot, K.B., and R.J. Mataloni. 2011. Operations of U.S. multinational companies in the United States and abroad. *Survey of Current Business* (November): 29–48.

Baumol, W.J. 2002. *The free-market innovation machine.* Princeton, NJ: Princeton University Press.

Bensen, Stanley M., and Leo J. Raskind. 1991. An introduction to law and economics of intellectual property. *The Journal of Economic Perspectives* (Winter): 23–24.

Bureau of Economic Analysis. 2011. U.S. International Services. http://www.bea.gov/international/international_services.htm#summaryandother.

Business Software Alliance and International Data Corporation (BSA/IDC). 2010. *09 Piracy Study.* Seventh Annual BSA/IDC Global Software Piracy Study. Washington, DC.

Capon, Noel, and Rashi Glazer. 1987. Marketing and technology: A strategic coalingment. *Journal of Marketing* 51 (July): 1–14.

Cohen, W.M., R.R. Nelson, and J.P. Walsh. 2000. Protecting their intellectual assets: Appropriablity conditions and why U.S. manufacturing firms patent (or not). Working chapter 7552. Cambridge, MA: National Bureau of Economic Research.

Corbet, J. 2007. Who Wrote 2.6.20? LWN.net, http:net/Articles/222773/.

Day, George S. 1994. The capabilities of market driven organizations. *Journal of Marketing* 58 (October): 37–52.

———. 1997. Maintaining the competitive edge: Creating and sustaining advantages in dynamic competitive environments. In *Wharton on dynamic competitive strategy*, ed. George S. Day and David J. Reibstein with Robert E. Gunther, pp. 48–75. New York: John Wiley & Sons.

Day, George S., and Robin Wensley. 1988. Assessing advantage: A framework for diagnosing competitive superiority. *Journal of Marketing* 52 (2) (April): 1–20.

Dixit, Avinash K., and Robert S. Pindyck. 1995. The options approach to capital investment. *Harvard Business Review* (May–June): 105–114.

Duysters, Geert. 1996. *The dynamics of technological innovation.* Brookfield, VT: Edward Elgar.

Economic Report of the President. 2011. *Gross Domestic Product.* Table B-1.

Ghemawat, Pankaj. 1986. Sustainable competitive advantage. *Harvard Business Review* 64 (5) (September–October): 53–59.

Glazer, Rashi. 1991. Marketing in an information intensive environment: Strategic implications of knowledge as an asset. *Journal of Marketing* 55 (October): 1–19.

Granstrand, Ove. 1999. *The economics and management of intellectual property.* Northampton, MA: Edward Elgar.

Griliches, Zvi, ed. 1984. *R&D, patents and productivity.* Chicago: The University of Chicago Press.

Grimaldi, Rosa, and Salvatore Torrisi. 2001. Codified-tacit and general-special knowledge in the division of labor among firms: A study of the software industry. *Research Policy* 30: 1425–1442.

Harabi, Najibe. 1995. Appropriability of technical innovations: An empirical analysis. *Research Policy* 24: 981–992.

Hillery, John S. 2004. Securitization of intellectual property: Recent trends. March. Washington/CORE. www.wcore.com.

Hulten, Charles R. 2010. Decoding Microsoft: Intangible capital as a source of company growth, NBER Working Paper 15799. Cambridge, MA: National Bureau of Economic Research.

Isckia, Thierry, and Dennis Lescop. 2009. Open innovation within business ecosystems: A tale from Amazon.com. *Communications & Strategies* 74 (2): 37–54.

John, George, Allen M. Weiss, and Shantanu Dutta. 1999. Marketing in technology-intensive markets: Toward a conceptual framework. *Journal of Marketing* 63 (special issue): 78–91.

Jones, C.I., and J.C. Williams. 1997. Measuring the social return to R&D. *Quarterly Journal of Economics* 113 (4): 1119–1135.

Koopmans T. 1957. *Three essays on the state of economic science.* New York: McGraw-Hill.

Kotler, Phillip, and Kevin Lane Keller. 2008. *Marketing management.* Upper Saddle River, NJ: Pearson Prentice Hall.

Kraemer, K.L., G. Linden, and J. Dedrick. 2011. Capturing value in global networks: Apple's iPad and iPhone. University of California, Irvine, University of California, Berkeley, Syracuse University, (July): unpublished paper 1–11. http://pcic.merage.uci.edu/papers/2011/Value_iPad_iPhone.pdf.

Mansfield, Edwin. 1985. How rapidly does new industrial technology leak out? *Journal of Industrial Economics* (December): 217.

———. 1995. *Intellectual property protection, direct investment, and technology transfer.* International Finance Corporation, Discussion Chapter Number 27. Washington, DC: World Bank.

Mansfield, Edwin, John Rapoport, Anthony Romeo, Edmund Villani, Samuel Wagner, and Frank Husic. 1997. *The production and application of the new industrial technology.* New York: W.W. Norton.

Mohr, Jakki. 2001. *Marketing of high-technology products and innovations.* Upper Saddle River, NJ: Prentice Hall.

Nadiri, M.J. 1993. Innovations and technological spillovers. NBER Working Paper No. 4423. Cambridge, MA: National Bureau of Economic Research.

National Science Board. 2012. R&D: National Trends and International Comparisons. *Science and Engineering Indicators 2012.* Chapter 4, Arlington, VA: National Science Foundation (NSB 12-01) (January).

Nordhaus, W.D. 1969. *Invention, growth and welfare.* Cambridge, MA: MIT Press.

———. 2004. Schumpeterian profits in the American economy: Theory and measurement. NBER Working Paper 10433. Cambridge, MA: National Bureau of Economic Research.

Peters, Thomas J. 1984. Strategy follows structure: Developing distinctive skills. *California Management Review* 26 (3) (Spring): 111–125.

Pharmaceutical Research and Manufacturers of America (PhRMA). 2010. *Pharmaceutical Industry Profile 2010.* Washington, DC: PhRMA.

Pisano, G.P. 2006. *Science business.* Boston: Harvard Business School Press.

Porter, Michael. 1980. *Competitive strategy: Techniques for analyzing industries and competitors.* New York: Free Press.

———. 1985. *Competitive advantage: Creating and sustaining superior performance.* New York: Free Press.

———. 1991. Towards a dynamic theory of strategy. *Strategic Management Journal* 12 (Winter): 95–118.

Porter, M.E., and Victor E. Miller. 1998. How information gives you competitive advantage. In *On competition*, ed. Michael E. Porter. Boston: Harvard Business School Press.

Rao, P.M. 1998. Intellectual property protection and foreign direct investment: A pharmaceutical industry perspective. In *Globalization, privatization, and the free market economy*, ed. C.P. Rao, pp. 232–253. Westport, CT: Quorum Books.

———. 1999. Convergence and unbundling of corporate R&D in telecommunications: Is software taking the helm? *Telecommunications Policy* 23: 83–93.

———. 2001a. The ICT revolution, internationalization of technological activity, and the emerging economies: Implications for global marketing. *International Business Review* 10: 571–596.

———. 2001b. International technology alliances: Recent trends in the I.T. sector. In *Globalization and its managerial implications*, ed. C.P. Rao, pp. 223–232. Westport, CT: Quorum Books.

———. 2012. "Monetization of intellectual property: An open innovation perspective." Proceedings of the International Conference on Technology Management, Indian Institute of Science, Bangalore. July 18–20.

Rao, P.M., and Joseph A. Klein. 1994. Growing importance of marketing strategies for the software industry. *Industrial Marketing Management* 23 (1): 29–37.

Rao, P.M., Joseph A. Klein, and Ramdas Chandra. 2011. Innovation without property rights and property rights without innovation: Recent development in the ICT sector. *Advances in Competitiveness Research* 19 (No. 1/2).

Richtel, Matt, and Jenna Wortham. 2011. Motorola's identity crisis. *New York Times*, www.nytimes.com/2011/08/22/technology.

Rohlfs, Jeffrey H. (2001). *Bandwagon effects in high technology industries*. Cambridge, MA: MIT Press.

Somaya, Deepak, David Teece, and Simon Wakeman. 2011. Innovation in multi-invention contexts: Mapping solutions to technological and intellectual property complexity, *California Management Review* 53 (4) (Summer): 47–79.

Schwarz, Michael, and Yuri Takhteyev. 2009. Half century of public software institutions: Open source as a solution to hold-up problem. NBER Working Paper 14946, *National Bureau of Economic Research*, Cambridge, MA.

Teece, David J. 1987. Profiting from technological innovation: Implications for integration, collaboration, licensing and public policy. In *The competitive challenge*, ed. David J. Teece, p. 185. Cambridge, MA: Ballinger.

———. 2000. *Managing intellectual capital*. Oxford: Oxford University Press.

Teece, David J., and Gary Pisano. 1994. The dynamic capabilities of firms: An introduction. *Industrial and Corporate Change* 3 (3): 537–556.

U.S. Census Bureau. 2012. *Corporations: Selected financial items: 1990–2008*. The 2012 Statistical Abstract.

U.S. Department of Justice (USDOJ). 1995. *Antitrust guidelines for the licensing of intellectual property*. Washington, DC: Government Printing Office.

U.S. Patent and Trademark Office (USPTO). 2005. Annual List of Top 10 Organizations Receiving Most U.S. Patents. (January): http://www.uspto.gov/news/pr/2005/05-03.jsp.

Vinod, H.D., and P.M. Rao. 2000. R&D and promotion in pharmaceuticals: A conceptual framework and empirical exploration. *Journal of Marketing Theory and Practice*, 8 (4): 10–20.

von Hippel, E. 2005. *Democratizing innovation*. Cambridge, MA: MIT Press.

West, Joel, and Scott Gallagher. 2006. Patterns of open innovation in open source software. In *Open innovation: Researching a new paradigm,* ed. Henry Chesbrough, Wim Vanhaverbeke, and Joel West, 82–106. Oxford: Oxford University Press.

West, Joel, Wim Vanhaverbeke, and Henry Chesbrough. 2006. Open innovation: A research agenda. In *Open innovation: Researching a new paradigm*, ed. Henry Chesbrough, Wim Vanhaverbeke, and Joel West, pp. 285–307. Oxford: Oxford University Press.

Williamson, Oliver E. 1975. *Markets and hierarchies: Analysis and antitrust implications*. New York: Free Press.

Winter, S.G. 1987. Knowledge and competence as strategic assets. In *The competitive challenge*, ed. David J. Teece, pp. 159–184. Cambridge, MA: Ballinger.

2 Analytical Tools for Marketing High-Tech

Chapter Overview

In this chapter, we discuss the traditional tools for analyzing strategic marketing issues and how they can be adapted to marketing high-tech products and services. Specifically, we discuss Porter's Five Forces model, Ansoff's growth strategy matrix, the Boston Consulting Group's (BCG) portfolio matrix, the experience curve, and the product life cycle. Porter's Five Forces model is a tool used to evaluate the relative profitability of an industry. We extend the discussion of this model to include the effects of globalization, the role of international organizations, and complements in the context of high-tech sectors. Ansoff's growth strategy matrix provides four broad strategies for growth. We describe those strategies and discuss their relevance to high-tech firms. The basic concept of the experience curve and its particular relevance to high-tech firms is explained. The chapter provides a discussion of BCG's portfolio matrix or the growth-share matrix in its simplest and extended forms, its relationship to the experience curve, and its limitations. Finally, a discussion of the familiar product life cycle, its value to firms operating in a high-tech environment, and its limitations is provided.

Porter's Five Forces Model

The theoretical underpinning of the "Five Forces" model is the structure-conduct-performance paradigm developed by Joe Bain (1956), an industrial organization (IO) economist, who suggested that the entry barriers contained within an industry's structure—economies of scale, absolute cost advantage, and product differentiation—determine the conduct (decisions) of buyers and sellers, and performance, in terms of innovation, productivity, and profitability, from a public policy perspective (Ghemawat 2006). Porter expanded on Bain's work and identified five forces that determine the degree of competition in an industry and its attractiveness in terms of profitability from a business strategy perspective (see Figure 2.1). These well-known forces include

Figure 2.1 **Porter's Five Forces Model for Industry Analysis**

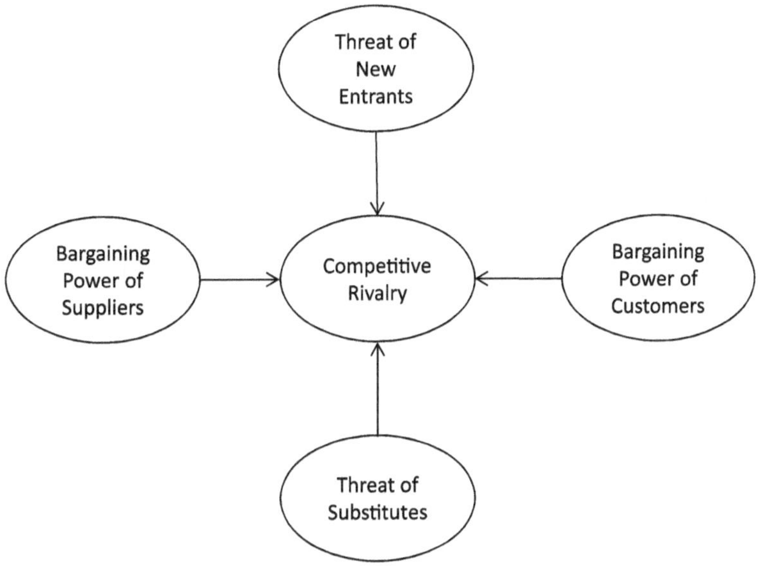

Note: For a detailed description of the Five Forces Model, see Porter (1980).

the threat of new entrants, threat of substitutes, power of buyers, power of suppliers, and competitive rivalry.

THREAT OF NEW ENTRANTS

The threat of new entrants depends on the relative importance of six major entry barriers and two others identified by Porter.

- Economies of scale
- Capital requirements
- Product differentiation (i.e., brand identity)
- Switching costs
- Access to distribution
- Cost disadvantages independent of scale
- Learning or experience curve effects
- Proprietary product technology
- Favorable locations
- Favorable access to raw materials
- Government subsidies
- Government policy
- Expected retaliation

In general, the presence of significant economies of scale, high capital requirements, product differentiation, significant proprietary learning curve effects, and in some cases switching costs—such as was the case with Windows operating system—represent relatively more important barriers to entry in high-tech markets. Moreover, in the context of high-tech entrants, discussed in chapter 1, a strong appropriability regime—determined by replicability that is hard and tight IP rights, combined with the complementary marketing assets that go beyond brand identity (e.g., relationship marketing)—becomes an important barrier to entry that favors investments and works against potential entrants, at least for a period of time.

Note, however, that new entrants with new technologies can weaken an otherwise strong appropriability regime. Thus, appropriability regime affects and is affected by the threat of new entrants. In addition, the significant depth and breadth in the technological activity of many high-tech firms, based on their R&D and patent profiles, serves as yet another barrier to entry. One need only consider the competition faced by traditional integrated pharmaceutical firms from the biotech firms in drug discovery or traditional facilities-based telecommunications firms from the Internet.

Government policy in the high-tech markets affects threat of entry in the form of threat of technical standards (as in the telecommunication industry), environmental standards (as in the chemical and auto industries), and safety standards (as in the pharmaceutical industry). Government-determined standards could affect the threat of entry either way. They could favor the incumbents as in the case of pharmaceuticals or favor the new entrants as happened in the telecommunications industry.

THREAT OF SUBSTITUTES

The threat of substitutes is determined by relative price performance of substitutes, switching costs, and buyer propensity to substitute. While all three factors determine the threat of substitutes in all industries, buyer propensity to substitute products—especially in the consumer goods area—with new technologies for ones with older technologies plays a greater role in high-tech industries. Typically, buyer propensity to substitute old products/technologies with new ones depends, other things being equal, on the willingness of customers to take risks with the substitutes. A small fraction of customers referred to as innovators take the risk, try the product, and, when satisfied, influence the buying propensity of imitators. There is vast literature on the diffusion of innovations, starting with the pioneering works by Everett Rogers (1962, 1983) and Frank Bass (1969), which will be discussed in chapter 6.

The threat of substitution arises not only from similar products and services (e.g., desktop and laptop computers) but also from products and services that are dissimilar but perform the same functions, as is the case, for example, with competition from the Internet and video conferencing to the growth of airlines, especially limiting the growth of the business travel segment. This type of substitution threat—that is, competition from physically dissimilar products and services—further distinguishes high-tech markets from other markets.

While a strong appropriability regime could reduce the threat of substitutes to some extent, substitutes created by new technologies with great speed (e.g., the substitution of VCR by DVD) weaken the appropriability regime. Thus, as is the case with the threat of entry, appropriability regime affects and is affected by the threat of substitutes, whatever the source of the threat may be.

POWER OF BUYERS

Porter identified several factors affecting the power of buyers:

- Substitute products
- Brand identity
- Buyer information
- Buyer concentration
- Switching costs
- Threat of backward integration
- Importance of quality/performance to the buyer
- Price sensitivity
- Product differentiation/brand identity
- Purchases from the industry as a fraction of total purchases
- Buyer profits

Even in high-tech markets, buyer power varies depending on whether the product is industrial or consumer, as well as many other factors. Take, for example, the commercial jet aircraft industry, which is dominated by two firms, Boeing and Airbus. This is an industry where threat of substitutes exists but is limited, at least in terms of direct substitutes. Buyer concentration (buyers are airlines) is not high enough to be a significant source of bargaining power, switching costs are high, ability to backward-integrate is low, product differentiation is significant, and some degree of brand preference is on the part of final consumers—the flying public—for Boeing versus Airbus planes for airlines to have much power.

On the other hand, in the telecom services industry, buyer power is significant because of the availability of cheaper substitutes from the Internet, little or no switching costs, weak brand identity, and so on. As a result, ability to appropriate value is lessened. This is despite the fact that buyer concentration and ability to backward-integrate are low.

POWER OF SUPPLIERS

Factors affecting supplier power—switching costs, supplier concentration, substitute inputs, and so forth—are essentially the same as the factors affecting buyer power, but from the perspective of suppliers. Power of suppliers in the context of high-tech markets is best examined in terms of the nature and extent of relationships between the firms in the industry and their input suppliers. Typically, high-tech firms operate with a high degree of vertical integration. (This is especially the case with R&D.) When firms outsource key inputs, they tend to form long-term relationships with suppliers in order to ensure quality and speed of development.

Collaborative arrangements such as joint ventures and alliances are also used to exchange and acquire technologies. In general, external sourcing of technology-related inputs leads to spillover of the firm's technology and weakening of the appropriability regime. Forward integration by suppliers using the technology gained from the buyers from their relationship is the ultimate risk.

COMPETITIVE RIVALRY

The four forces discussed above—competition from new entrants, substitutes, the power of buyers, and the power of sellers—influence the degree of rivalry in an industry. Porter's determinants of rivalry are:

- Industry concentration
- Industry growth
- Capacity utilization
- Fixed costs
- Product differentiation/brand identity
- Switching costs
- Diversity of competitors
- Strategic stakes
- Exit barriers

Note that several of the determinants of rivalry overlap with the factors influencing the other four forces as well. Switching costs, which are particularly important in many high-tech markets, like computers and telecommunications, show up in all five forces. The high switching costs of retraining, product redesign, and so on serve as effective barriers to entry, reduce the power of buyers, and hence lower the degree of rivalry. While a high degree of industry concentration, rapid growth, high degree of capacity utilization, high fixed costs, and significant product differentiation and brand identity serve as effective barriers to entry and therefore reduce rivalry in all industries, strategic stakes, which involve firms making large bets on a certain technology, are particularly high in the high-tech markets. Examples of strategic stakes range from IBM's bet on the success of open source software in competition with Microsoft's proprietary software to Airbus's bet on the $485 million super jet A380 against Boeing's 787 Dreamliner. Rivalry in industries characterized by such high strategic stakes is likely to be intense and unstable until the outcome is settled one way or another.

BEYOND THE FIVE FORCES MODEL

Globalization

Since the appearance of Porter's Five Forces model three decades ago, the structure of the United States and other industrialized economies, along with the nature of competition, has changed greatly. By almost any measure, firms now compete in a vaster and more globalized world. The service sector dominates these economies. As noted earlier, increasingly intangible assets in the form of intellectual and marketing capital define these economies. Aided by the information and communications technology (ICT) revolution, multinationals are increasingly internationalizing even their R&D activity. Technology alliances and joint ventures between firms (often between competing firms) and networks became commonplace, notwithstanding the potential loss of proprietary knowledge. Competition and cooperation coexist. Nongovernmental and nonprofit organizations such as the Bill & Melinda Gates Foundation have become significant global players in important high-tech industries such as pharmaceuticals.

Moreover, although government policy in the Five Forces model has been discussed mainly in terms of its impact on entry barriers, Porter (1980) recognized how government regulations, enforcement of antitrust laws, subsidies, taxes, and other means can influence the power of buyers, suppliers, the threat of substitutes, and ultimately the degree of rivalry within an industry. Indeed, Porter himself suggested that "strategy may well involve treating government as an actor to be influenced" (p. 29). The trend has been toward cooperation and partnership between government and industry. This is especially the case with high-tech industries (aerospace, for example), viewed as "strategic" from the standpoint of national competitiveness.

Role of International Organizations

Structural analysis of industries also needs to consider the vastly increased role of international organizations and compacts such as the World Trade Organization and numerous regional trade associations like the European Union and the North American Free Trade Agreement, or NAFTA. As will be discussed in a later chapter, nowhere was government-industry cooperation more evident than in the negotiations between the developed countries and the developing countries that led to the Agreement on the Trade Related Intellectual Property Rights (TRIPs) in the Urgency Round. The TRIPs agreement was intended to strengthen the appropriability regime faced by the developed countries' multinationals in the developing countries where intellectual property rights (IPRs) were loose or nonexistent. The agreement is particularly important to high-tech sectors like pharmaceuticals and software. So the relative strength of the appropriability regime determined by the IPRs, replicability of knowledge, and complementary marketing assets becomes yet another important explicit variable affecting all five forces.

Complements

Another important extension of the Five Forces model that is particularly important for the high-tech industry was suggested by Ghemawat (2006). Drawing upon the value net framework developed by Brandenburger and Nalebuff (1996), Ghemawat emphasized the role of complementors, that is, firms from whom customers buy complementary products and services. For example, an independent software vendor that writes applications software for the Windows operating system increases the willingness of customers to pay for Windows. Ghemawat adds such complementors as a sixth force to Five Forces framework and emphasizes cooperation with complementors to expand the size of the market without losing sight of potential competition with them. Our approach in this book is to adapt the Five Forces model and other tools discussed in this chapter to the strategic analysis of high-tech industries and firms.

ANSOFF'S GROWTH STRATEGY MATRIX

While Porter's Five Forces model is typically used at the industry level, other strategic planning tools are used at the firm level. In the traditional growth strategy matrix, also known as product-market grid, proposed by Ansoff (1965) (see Figure 2.2), a firm first considers growing through greater market share in its current market with its current products. This is known as *market penetration strategy*. Next, it considers growth through entering new markets—domestic and/or international—for its current products, which is known as *market development strategy*. New markets can be new segments of the existing market. Then it considers developing new products for its current markets (*product development strategy*) followed by *diversification strategy*, which includes developing new products for new markets. Note that others (Johnson and Jones 1957, for example) have modified and expanded the Ansoff's matrix by classifying new products in terms of technological newness (e.g., improved technology versus new technology) and market newness (e.g., strengthened market versus new market). More recently, Aaker (2005) supplemented Ansoff's classification of growth opportunities by including vertical integration (i.e., contribution of the firm to the value chain); and Strebel and Ohlsson (2006) along the dimensions of innovation and other variables. The reader is asked to consult Roberts (2012) for a detailed discussion of extensions to the Ansoff matrix.

DIVERSIFICATION STRATEGY

Diversification into new products or services has four forms—horizontal, vertical, concentric, and conglomerate.

Horizontal Diversification

This occurs when a given firm adds new products or services through acquisition or internal development that are technologically or commercially unrelated to its current products or services, but that may appeal to current customers. Canon, known for its cameras, moving to desktop copiers would be an example of horizontal diversification.

Vertical Diversification

This occurs when a company, through backward integration, moves into the business of its input supplier (JPMorgan Chase decides to move into software development), or when a company, through forward integration, moves into the business of its customers (Intel moves into production and marketing personal of computers).

Figure 2.2 **Ansoff's Growth Strategy Matrix**

Product Offering \ Market	Current	New
Current	**Market Penetration Strategy**	**Market Development/Expansion Strategy**
New	**Product Development/Expansion Strategy**	**Diversification Strategy**

Source: Adapted from Ansoff (1965), Figure 6.1.

Concentric Diversification

This takes place when new product or service areas into which a company is diversifying are related to the existing product lines in terms of technology and/or marketing. General Electric, a leader in the production and marketing of artificial industrial diamonds, produces and markets artificial diamonds for the consumer market. That would be a case of concentric technological diversification.

Conglomerate Diversification

This means that the companies' new products or services have neither technological nor marketing synergies with its current business. The objective is to provide increased growth and profitability through a well-diversified and well-managed portfolio of strategic business units. General Electric, with its diversified business ranging from light bulbs to jet engines, broadcasting, and finance, fits the description of a conglomerate.

RELATIVE RISKS OF GROWTH STRATEGIES

In terms of relative risks, diversification strategy—new products or services in new markets—is perhaps the riskiest of growth strategies. Concentric diversification, which is intended to exploit synergies in technological and/or marketing activities of the firm, mitigates some of the risks of diversification. Conglomerate diversification is perhaps the riskiest of all diversification strategies. Horizontal and vertical diversification strategies fall between conglomerate and concentric diversification in terms of risk. Market penetration strategy—current products or services in current markets—is the least risky of the growth strategies, with product development and market development strategies falling in between the diversification and penetration strategies.

GROWTH STRATEGY MATRIX AND HIGH-TECH FIRMS

Ansoff's growth strategy matrix is generic and therefore applicable to many products and services, including high-tech ones. However, several factors distinguish growth strategies for high-tech firms from other firms. First, high-tech firms, in contrast to other firms, depend on a continuous stream of new products and services for their growth. Although like other firms, high-tech firms engage in acquisitions for growth, a history of successful in-house development of new products and services is their raison d'être. This is true of firms in the software industry, the prescription drug segment of the pharmaceutical industry, and many other high-tech sectors. The focus is on new product development requiring large up-front risky investments in R&D. Second, the failure rate of new products is rather high. Overall, the failure rate of new products is close to 40 percent (Day 2007). Nine out of ten prescription drugs in the pharmaceutical sector fail, for example. So, firms need to pursue a strategy of concentric diversification into related technologies. That is, technological diversification precedes product diversification. Note that what promising products and services come out of this technological diversification is not easily predictable. Third, given the high degree of technical and market uncertainty associated with high-tech products and services, firms would have difficulty moving, sequentially, from a market penetration strategy to a diversification strategy implied by the Ansoff matrix. Serious threat of entry combined with a weak appropriability regime calls for aggressive pursuit of penetration strategy. This may require simultaneous implementation of market penetration and market development strategies.

INNOVATION PORTFOLIOS

Finally, as Day (2007) suggests, while minor innovations make up 85 percent to 90 percent of companies' innovation portfolio, they do not help achieve growth objectives. It is the big, albeit risky, projects that contribute disproportionately to growth and profitability. According to a study by Kim and Mauborgne (1999), only 14 percent of new product launches were significant innovations, but they accounted for 61 percent of the profits from innovations. Therefore, it is important to have a balanced innovation portfolio in terms of risk and reward.

Risk Matrix and R-W-W Questions

Day (2007) suggests two tools to help the companies accomplish this balance. One is the risk matrix, which shows the risk across a company's innovation portfolio (see Figure 2.3) and the other is R-W-W ("real, win, worth it"), a tool to evaluate the risks

Figure 2.3 **Risk and Innovation Portfolio**

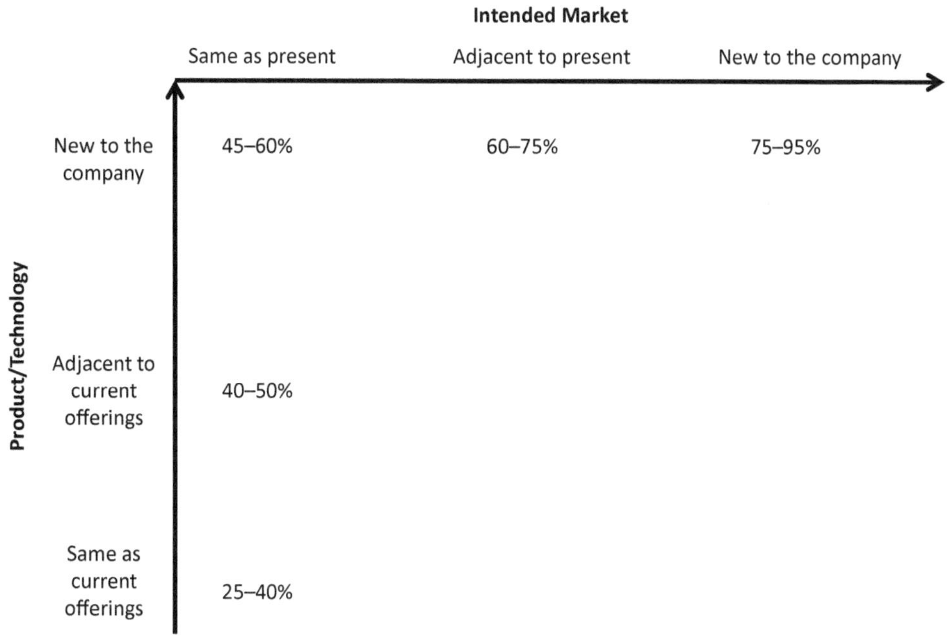

Source: Adapted from Day (2007).
Note: The percentages represent probability of failure.

and potential of projects by answering three questions. Potential market and the feasibility of developing the product are determined by the answer to the questions "Is it real?" and "Can we win?" The answer to these questions considers the competitive advantage of the innovation and the firm. The profit potential and strategic fit are assessed by the third question: "Is it worth doing?"

EXPERIENCE CURVE

Another strategic tool is the experience curve concept, a broader version of the learning curve concept, which suggests that labor cost per unit in real terms declines by a constant percentage as cumulative output doubles. This phenomenon was first observed in the aircraft industry in the 1930s (Wright 1936). Some forty years later, the Boston Consulting Group (BCG) expanded the learning curve concept and called it the experience curve, which considered the effect of learning-by-doing on labor costs associated with production as well as marketing, administrative, and other overhead costs. The BCG and others quantified this phenomenon across a broad range of products and found that real cost per unit did decline by a predictable percentage—ranging from 10 percent to 30 percent—with each doubling of output, a proxy for doubling "experience" in delivering a product or service.

Figure 2.4 illustrates a "70 percent experience curve"—plotted on log-log scale, as a straight line—which shows that the real cost per unit declines by 30 percent with each doubling of cumulative experience (Figure 2.5 shows the same "70 percent experience

Figure 2.4 **The 70 Percent Experience Curve** (log-log scale)

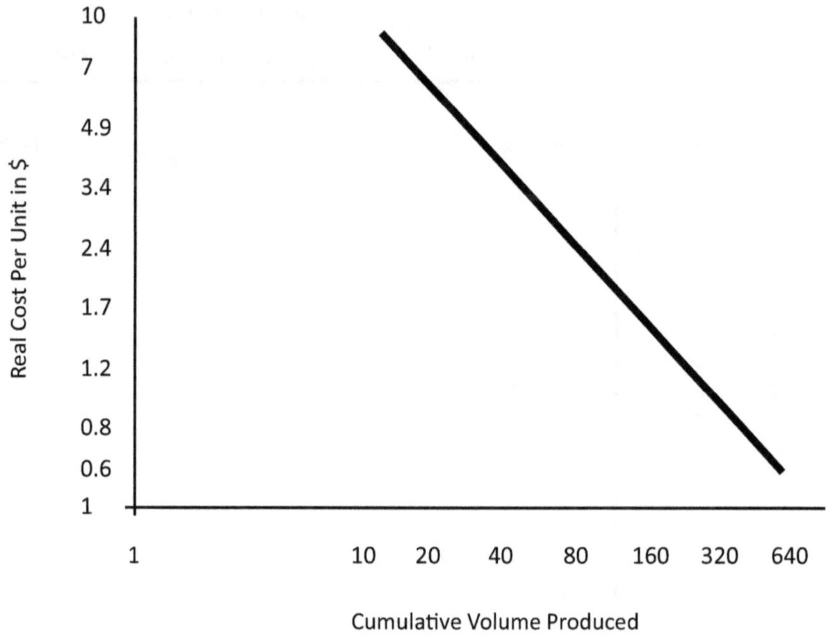

Figure 2.5 **The 70 Percent Experience Curve** (arithmetic scale)

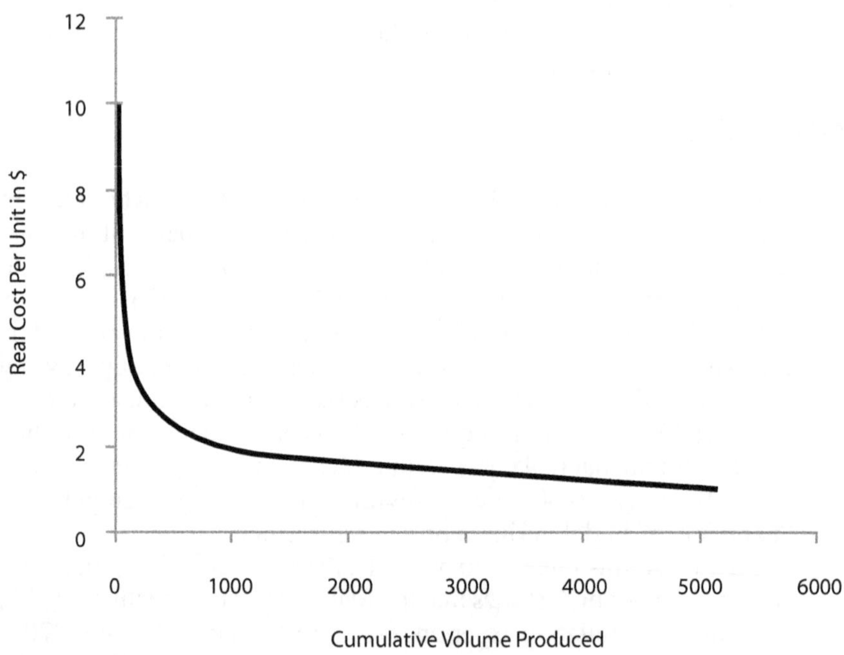

curve" in arithmetic scale). Note that the experience curve in Figure 2.5 reflects the aggregate effect of experience on costs. Individual elements that comprise the aggregate curve such as R&D, marketing, and administrative overhead have their own experience curves. Sources of cost reductions are many and go beyond learning-by-doing. Economies of scale are an important source of cost reductions because capital investment and many overhead costs such as R&D, marketing, and administration do not double when output doubles. Improvements in product and process technology driven by the volume of production are another source of experience curve effect. Therefore, the slope of the experience curve depends on the opportunity—and the ability of the firm—to exploit these various sources.

EXPERIENCE CURVES IN THE HIGH-TECH CONTEXT

In the context of high-tech products and services, the relative newness of technology and its complexity, combined with the willingness of the firm to invest necessary resources, affords great opportunity to exploit the experience curve effects. Economies of scale and economies of scope, or economies of doing things under "one roof" in R&D, would be significant sources of experience curve effects in high-tech firms. Improvements in product design in the form of product standardization process and assembly-line type process—both of which drive the volume and are driven by the volume—are other major sources of experience curve effect in high-tech firms.

Researchers observed significant experience curve effects for many high-tech products—integrated circuits (70 percent curve); 3-ton spilt system air conditioners (80 percent curve); steam turbine generators (70 percent curve); and photovoltaic technology (80 percent curve). The strategic implication of the experience curve phenomenon is clear. A firm that enters the market and has the opportunity and the resources to exploit the experience curve has lower costs, achieves rapid growth, and captures a significant share of the market, which, in turn, leads to still lower costs and higher market share. In short, the message of the experience curve is to gain competitive advantage through lower costs, lower prices, rapid growth, and dominant share of the market. This, indeed, is the conceptual underpinning of the growth/share matrix (also known as the portfolio matrix or the BCG matrix) developed and popularized by the Boston Consulting Group, discussed in the next section. While low-cost strategies based on the experience curve can limit entry and provide significant advantage, as Czepiel (1992) notes, they run up against the market's demand for product variety and the competitor's ability to produce variety through new technologies, which have their own experience curve effects.

GROWTH-SHARE MATRIX

As noted earlier, the experience curve—with its emphasis on growth and market share—provides the conceptual foundation for the growth-share matrix. As shown in Figure 2.6, the two-by-two matrix is based on the real growth rate of the relevant industry/sector and relative market share, that is, relative to that of the largest competitor in the industry. The basic idea behind the growth-share matrix, also known as

Figure 2.6 **Growth-Share Matrix and Its Relationship to Product Life Cycle**

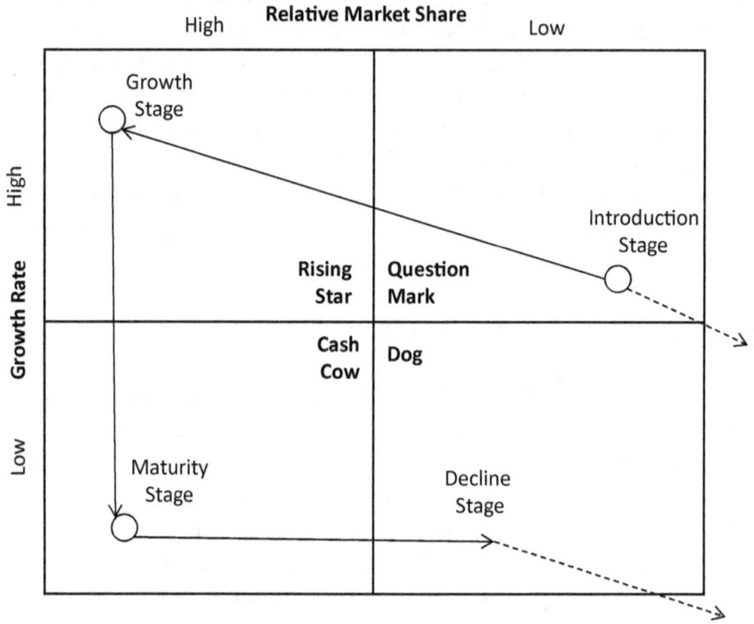

the BCG matrix, is that the firm exploits the experience curve effects (presumably by being first to introduce the product) and operates at a relatively low cost/price position, which produces rapid growth and market-share dominance. Rising stars are business units/products in high-growth markets with relatively high markets share. They are large cash users as well as cash generators. Cash cows are business units/products in low-growth markets with high relative share. They generate more cash than they can use, so they can support promising question marks—business units/products in fast-growing markets with low relative share—and some stars. Dogs—businesses/products with low relative share in a slow-growing market—are to be phased out. For the origin of the BCG matrix see Henderson (1970). The obvious relationship between growth-share matrix and the product life cycle (PLC)—moving counter-clockwise from question marks (introductory stage in the PLC), stars (growth stage), cash cows (maturity stage), and dogs (declining stage), which will be discussed in the next section—is also shown in Figure 2.6.

Limitations of the Growth-Share Matrix

Several limitations of the growth-share matrix have been noted by Porter (1980), Czepiel (1992), Ghemawat (2006), and others. They include the following: The relative share is often not a good proxy for competitive position, and relative costs and share itself may be a result of other factors such as superior management; industry growth rate (as the Five Forces model suggests) is only one of many factors that determines the profitability of a business in that industry; excessive emphasis on internal balancing of negative cash flows from question marks and some rising stars with positive cash flows from cash cows, given the availability of external financing for financially worthy projects; strategic decisions cannot be made on just two dimensions; synergies between business units are neglected, in part, because of the difficulty of defining the market; and that the empirical evidence on the relationship between market share and profitability (return on investment) is weak, at best.

GE/McKinsey Matrix

Despite these limitations, the growth-share matrix could be useful to large high-tech firms that can exploit experience curve effects—when they are present—through volume if it is modified and used in conjunction with other tools. One such modification came from the Industry Attractiveness–Business Strength Matrix developed by GE and McKinsey and Company. The GE/McKinsey matrix screens business strength and industry attractiveness by about a dozen measures each. For example, business strength includes leadership in technology, manufacturing, marketing and distribution, and product quality, in addition to relative market share and profitability. Industry attractiveness includes market size, profit margin, market diversity, legal and environmental factors, in addition to growth rate. Technology development is a factor that affects industry attractiveness and a record of innovation is a factor that affects the strength of the business unit. For high-tech firms the nature of appropriability regime can be another factor that affects industry attractiveness. Note, however, that the GE/McKinsey matrix, because of large numbers of factors—some easily quantifiable and others not—and despite attempts to weight them, could, unlike the growth-share matrix, be very subjective as far as where a business unit falls in the matrix. Moreover, the GE/McKinsey matrix, like the growth-share matrix, does not take account of synergies—particularly synergies in R&D, which are particularly important for high-tech firms. Such synergies are known as economies of scope or economies that result from performing several related projects under one roof. The reader is referred to John A. Czepiel (1992) for a detailed discussion of the GE/McKinsey matrix.

Product Life Cycle

The familiar product life cycle shown in Figure 2.7 assumes that most products follow an S-shaped curve. The shape is determined by the process of innovation and diffusion of new products, and each new product goes through a number of stages—introduction, growth, maturity, and decline—and the stages are defined by the inflection points in the S-curve. Profits are low or negative in the introductory stage because of the low level of sales combined with high marketing costs associated with launching a new product, and highest in the growth stage. It is recognized that the decline of the product may be delayed through market penetration, product development, and market development strategies discussed earlier (see Figure 2.2). Recall, also, the relationship of PLC to the growth-share matrix. Since every new product is a question mark, it is associated with the introductory stage. Stars are associated with the growth stage, cash cows with maturity stage, and dogs with the declining stage.

Figure 2.7 **Product Life Cycle**

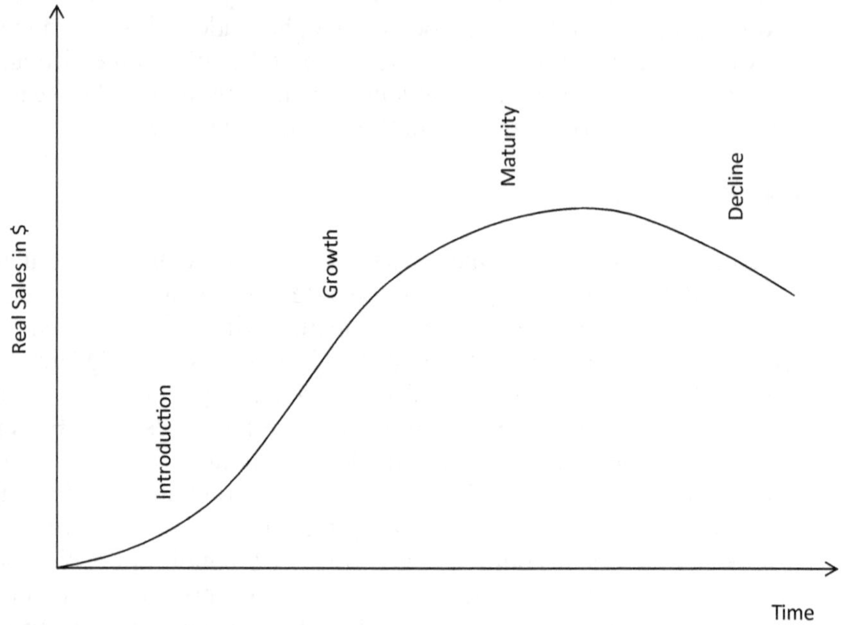

For purposes of developing functional marketing strategies (i.e., product, pricing, promotion, and distribution strategies) each stage of the PLC is best viewed as an adopter segment in terms of Rogers's (1983) *Diffusion of Innovations*. Innovators (2.5 percent) comprise the introductory stage; early adopters (13.5 percent) and early majority (34 percent) are associated with early and later part of the growth stage, respectively; late majority (34 percent) with the maturity stage; and laggards (16 percent) with the late maturity and decline stage.

Despite its popularity, the PLC concept has attracted criticism on several grounds. Not all products go through the S-curve. A single variable, time, explains the shape of the S-curve. Competitive forces associated with each stage of the PLC vary across industries. This is particularly the case with respect to high-tech industries. The market structure, technology, and the underlying competitive forces influencing the shape, the stages, and their duration are very different for high-definition television (HD-TV), for example, compared with the CD player, whose diffusion rate is faster and life shorter because of convergent technologies like the DVD, iPhone, and digital television and penetration pricing.

Prior to launching a new product, a firm needs to make two key strategic decisions. One is the timing of the entry (enter as a pioneer? as a follower?) and the other is mode of entry (internal development versus other modes such as joint venture, alliance, or acquisition). Both decisions are influenced by the nature of appropriability regime. Pioneer strategy is a high-risk and high-return strategy. Pioneers make large up-front investments in R&D to develop a new product, which entails technological as well as commercial risks. Risk of imitation and leap-frogging by followers is always present. Against these risks, the pioneer needs to create first-mover advantages through experience curve effects, branding, head-start on distribution that is relationship-based, continuous improvement in technology, economies of scale and scope in R&D, dominant product design that becomes an industry standard, presence of significant switching costs, and the like that act as barriers to entry. These advantages are supplemented by penetration strategy, that is, squeezing more sales out of current product and current market through greater usage among current users and finding new segments for the current product. Pricing strategy also plays an important role in augmenting the competitive advantage created by the pioneer firm. The more the advantages, the greater is the likelihood that the pioneer firm perceives potential competition as low and follows a skimming strategy. That is, the firm starts with a high initial price in the introduction stage and gradually reduces the price over the PLC, a strategy that is discussed at length in chapter 3.

A follower strategy, on the other hand, is based on waiting until the customers' needs are better established, letting the pioneer take the early risks, learning from the pioneer's mistakes, and coming up with a product that is far superior to the pioneer's product and serving the same segments (e.g., introduction of iPod by Apple in competition with the MP3 player designed by Archos Jukebox). Alternatively, a follower could leapfrog the pioneer with major changes in technology and design that attempt to directly challenge the pioneer's competitive advantage, such as the Nokia Lumia 900, which has Microsoft's Windows Phone 7.5 as the operating system attempting to leapfrog Apple's stronghold over the smartphone market.

INNOVATION CYCLE VERSUS PRODUCT LIFE CYCLE

The typical textbook model of the PLC discussed above and shown in Figure 2.7 assumes, among other things, the emergence of a dominant design and proceeds to analyze its implications for marketing strategy over the four stages. However, this is not the case, as Utterback (1994) has shown, in terms of the dynamics of innovation cycle (IC). According to Utterback, IC goes through three phases—the fluid phase, the transitional phase, and the specific phase—as shown in Figure 2.8.

The fluid phase is characterized by frequent and significant product changes, with no dominant design; R&D is still not significantly focused because of a high degree of technical uncertainty and a fast growing number of competitors are challenging the pioneers. The basis of competition is functional performance of the product. The transitional phase requires market acceptance of a product innovation and the emergence of a dominant design. Without them there is no growth for a new product and there is no transitional phase. The focus of innovation in the transitional phase shifts to process technology to meet the growing demand and the need to improve efficiency and quality. R&D focuses on specific product features, and firms compete on the basis of product variety. During the specific phase, the focus of innovation (and R&D) is on incremental improvements to standardized products; emphasis on process technology; competitors are few with stable market shares; the basis of competition is price; and pioneers are vulnerable to superior products substitutes.

Figure 2.8 **Innovation Life Cycle**

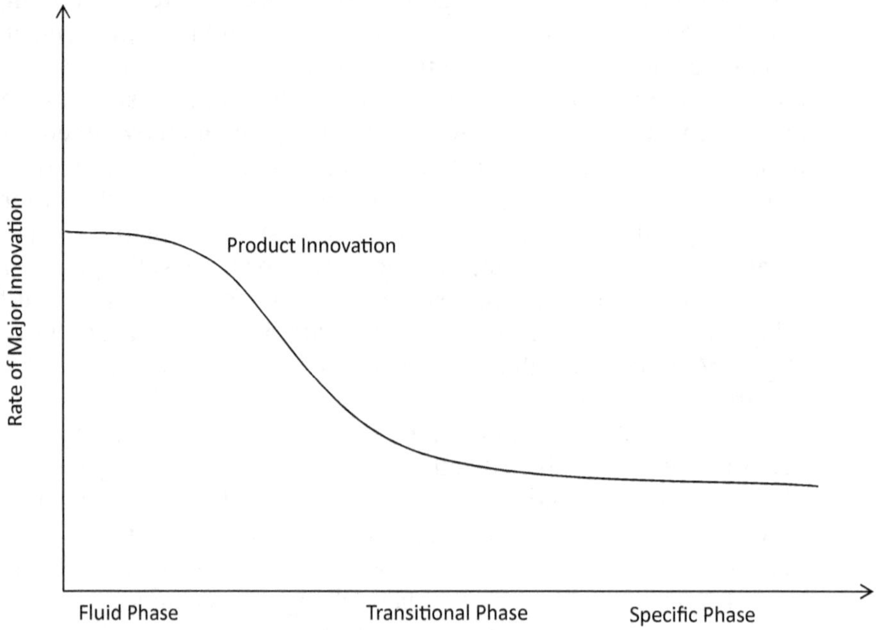

Source: Adapted from Utterback (1994).

In terms of concordance between the classic PLC in Figure 2.7 and the innovation cycle (IC) in Figure 2.8, the fluid phase of the IC corresponds roughly to the introductory stage (especially the early part) of the PLC; the transitional phase of the IC with the later part of the introductory stage and the growth stage of the PLC; and the specific phase with the maturity stage of the PLC. The greatest challenge for the marketing strategy is in the fluid phase of the IC because of high degree of market and technical uncertainty. This is also the phase in which pioneers—with or without intellectual property rights—are most vulnerable to imitators. The focus of marketing strategy is to gain quick consumer acceptance in the face of competing designs and uncertain standards. Communication strategy focused on early adopters becomes central to the success of new product. Equally important is the marketing–R&D interaction in terms of design changes based on customer feedback and competitive intelligence.

SUMMARY

Five standard analytical tools with relevance to high-tech firms' strategies have been discussed. In Porter's Five Forces model, the degree of competition in an industry and its profitability are determined by the threat of entry by competitors, the threat of substitutes, power buyers, and suppliers, all of which together influence the degree of competitive rivalry. The presence of scale economies, experience curve effects, switching costs, and strong brand identity, among others, reduces the threat of new entrants and competitive rivalry, and increases, ceteris paribus, the relative profitability of an industry. Appropriability affects and is affected by the threat of entry. Similarly, a strong appropriability regime could reduce the threat of substitutes created by new technologies with great speed (e.g., DVD replacing VCR) and weaken the appropriability regime. An industry's profitability is also determined by the relative power of suppliers and buyers, which in turn is determined by switching costs, supplier and buyer concentration, and the threat of forward integration by the suppliers and backward integration by the buyers.

Although the Five Forces model remains one of Porter's important contributions to the strategy literature, a lot has changed since its appearance some three decades ago. The boundaries of industries are greatly blurred. This is especially the case with the high-tech ICT sector comprising telecommunications, computers, and software. Competition and cooperation coexist between firms and industries. Nonprofit organizations like the Bill & Melinda Gates Foundation have become significant players in important high-tech industries such as pharmaceuticals. Cooperation and partnership between government and industry, especially in the high-tech sectors, has increased to the point where treating government as an actor to be influenced is widely recognized. Similarly, the vastly increased role of international organizations such as the WTO and regional trade associations in the world's economies needs to be considered in any structural analysis of industries. The growing importance of complementors (firms and individuals that specialize in writing applications software and help expand the size of the market, for example) is another trend that influences the structure and profitability of industries.

The riskiest of the four growth strategies in the Ansoff growth strategy matrix is the diversification strategy involving the development of new products as well as new markets, which is the focus of high-tech firms. That said, high-tech firms cannot achieve sustained growth and profitability unless their innovation portfolio includes big, albeit risky, projects.

Understanding and exploiting the experience curve concept is particularly beneficial to high-tech firms. The basic idea of the experience curve is that real unit costs decline by a fixed percentage every time production/real sales double due to learning-by-doing. The concept applies to all costs, including marketing costs. This phenomenon creates strategic advantage to firms that enter the market first. It gives the firm flexibility in pricing to quickly gain market share and profitability. The experience curve, with its emphasis on growth and market share, provides the conceptual foundation for the growth-share matrix.

With respect to product life cycle as a tool of analysis, it should be noted that its application to high-tech is limited because the market structure, technology, and underlying competitive forces influencing the shape, the stages, and the duration vary widely across industries and products. How a firm can use the PLC concept to its strategic advantage depends on the timing of entry (pioneer versus follower) as well as the mode of entry (internal development versus joint venture). Both of these decisions are influenced, in turn, by the nature of the appropriability regime faced by the firm.

KEY TERMS

backward integration
concentric diversification
conglomerate diversification
economies of scale
economies of scope
horizontal diversification
innovation cycle
innovation portfolio
product life cycle
risk matrix
switching costs
vertical diversification

DISCUSSION QUESTIONS

1. (a) Describe Porter's "Five Forces" model of industry competition and identify the key elements in this model that are particularly relevant to high tech-industries.
   (b) What are the missing elements in the model in light of the current structure of the world economy?
2. Describe the four generic growth strategies developed by Ansoff and discuss the factors that distinguish growth strategies for high-tech firms versus other firms.
3. Discuss the two tools developed by Day—risk matrix and R-W-W—to help firms achieve a balanced innovation portfolio.
4. (a) Explain the basic idea of the experience curve and discuss why it is particularly relevant to high-tech firms.
   (b) What is the relationship between experience curve effects and the growth-share matrix?
5. Discuss critically the concept of product life cycle and innovation cycle and their relevance to marketing strategy in high-tech firms.

REFERENCES

Aaker, D. 2005. *Strategic market management.* 7th ed. Hoboken, NJ: John Wiley & Sons.

Ansoff, Igor H. 1965. *Corporate strategy: An analytic approach to business policy for growth and expansion.* New York: McGraw-Hill.

Bain, Joe S. 1956. *Barriers to new competition.* Cambridge, MA: Harvard University Press.

Bass, Frank M. 1969. A new product growth model for consumer durables. *Management Science* (January): 215–227.

Brandenburger, Adam, and Barry Nalebuff. 1996. *Co-opetition.* New York: Doubleday.

Czepiel, John A. 1992. *Competitive marketing strategy.* Englewood Cliffs, NJ: Prentice Hall.

Day, George S. 2007. Is it real? Can we win? Is it worth doing? Managing risk and reward in an innovation portfolio. *Harvard Business Review* (December): 110–120.

Ghemawat, Pankaj. 2006. *Strategy and the business landscape.* Upper Saddle River, NJ: Pearson Prentice Hall.

Henderson, Bruce. 1970. *The product portfolio,* Boston: Boston Consulting Group.

Johnson, Samuel C., and Conrad Jones. (1957). How to organize for new products. *Harvard Business Review* 35: (May–June) 49–62.

Kim, W. Chan, and Renée Mauborgne. 1999. Strategy, value innovation, and the knowledge economy. *Sloan Management Review* (Spring): 41–54.

Porter, Michael E. 1980. *Competitive advantage: Techniques for analyzing industries and competitors.* New York: Free Press.

Roberts, John M. 2012. New product development in a strategic context. In *Handbook of marketing strategy,* ed. Shankar Venkatesh and Gregory S. Carpenter, 172–197. Cheltenham, UK; Northampton, MA: Edward Elgar.

Rogers, Everett M. 1962. *Diffusion of innovations*. New York: Free Press.

———. 1983. *Diffusion of innovations.* New York: Free Press.

Strebel, P., and A.V. Ohlsson. 2006. The art of making smart moves. *Sloan Management Review* 47 (2) (Winter): 79–83.

Utterback, James M. 1994. Innovation and industrial evolution. In *Mastering the Dynamics of Innovation*. Boston, MA: Harvard Business School Press.

Wright, T.P. 1936. Factors affecting the cost of airplanes. *Journal of the Aeronautical Sciences* 3 (1996): 122–128.

3 Pricing Strategies for High-Tech Products and Services

Chapter Overview

This chapter discusses how several distinguishing features of high-tech products and services guide their pricing strategies. The chapter then provides a detailed discussion of examples of pricing strategies from three high-tech sectors—pharmaceuticals, commercial jet aircraft, and software. In addition, the chapter discusses the concept of disguised price discrimination, bundling, experience curve pricing, penetration pricing, bandwagon effects, and pricing of technology.

Introduction

Pricing strategies for high-tech products and services are guided by several features of such products and services that distinguish them from other products. Below, the differences are highlighted.

First, high-tech products and services are highly differentiated. A high degree of R&D intensity often accompanied by a high level of complementary promotional intensity is an indicator of a high degree of differentiation. Differentiation provides a degree of control over prices for the firm. That is, the firm becomes a price setter (Apple, for example) rather than a price taker (a wheat farmer in Kansas, for example). The degree to which firms have control over prices depends on the market structure in which they operate. Price setting firms are said to operate in a market structure of imperfect competition and price taker firms are said to operate in a perfectly competitive market. In a market structure of imperfect competition, firms engage in nonprice competition based on quality, service, brand name, and other qualitative factors.

In a perfectly competitive market, firms compete on the basis of price. When high-tech firms compete on the basis of quality and brand name, they enter into an R&D and promotion race in which only a small number of firms survive and the threat of new entrants is low. Economists classify imperfectly competitive markets into three broad market structures: monopolistic competition (many producers with real or apparent product differentiation, e.g., cell phones and personal computers), oligopoly (few producers with varying degrees of product differentiation, e.g., autos and commercial jet aircraft at the high end of differentiation and aluminum and steel at the low end), and monopoly (single producer with no real substitutes, e.g., patented drugs and Windows operating system). Producers in each of these market structures face downward sloping demand curves with varying degrees of control over prices—of the greatest degree in monopoly—achieved through investment in R&D and promotion. Another way to think about investments in R&D and promotion is that they are aimed at creating a market structure favorable to the firm, namely a monopoly or a near monopoly.

Second, both R&D and promotional expenditures, which dominate high-tech firms, are sunk costs. That is, costs that are irreversibly committed and do not factor into present pricing decisions. This does not mean R&D and promotion costs are ignored and not recovered. All costs are important and must be covered if the firm is to survive. The point is that profit-maximizing prices are those that maximize contribution margin, that is, the difference between marginal (incremental) revenue and marginal (incremental) cost.

Third, many high-tech products and services exhibit experience curve effects. That is, as shown in chapter 2, cost per unit declines by a fixed percentage every time production doubles. This has important implications for pricing strategy.

Fourth, as noted in chapter 1, pricing strategies of high-tech firms over the product life cycle depend in part on the nature of appropriability regime and IPRs.

Fifth, digital high-tech products such as CDs, DVDs, and other information products have high costs of production and near-zero cost of reproduction, conditions that present special challenge in terms of pricing strategies. The rest of this chapter is devoted to discussion of these topics.

PRICING STRATEGIES IN IMPERFECTLY COMPETITIVE MARKETS

In this section, we discuss examples of pricing strategies in three different market structures and industries—the case of temporary monopoly in pharmaceuticals, the case of oligopoly in the commercial jet aircraft industry, and the case of monopolistic competition in the applications segment of the computer software industry.

THE CASE OF TEMPORARY MONOPOLY IN PHARMACEUTICALS

Firms specializing in the production and marketing of prescription drugs enjoy a temporary monopoly through strong patent protection in the United States and most other industrialized countries. Statutory patent life is twenty years and effective patent life is about ten years. R&D and promotion intensities of these firms are among the highest in the United States. Firms gain physician acceptance of a drug and build strong brand name during the patent period by investing heavily in promotional expenditures—well over 40 percent of sales. Generally, close substitutes for a successfully launched patented prescription drug are few and far between, but they exist. However, post-patent cut-throat price competition from domestic and foreign producers of generic drugs—who spend little or nothing on R&D—is expected. So, innovative firms need to consider potential price competition in designing pricing strategies. On the average, demand for prescription drugs is relatively inelastic, considering especially that in the United States the majority of the population is covered by insurance.

Given the background above, how do the producers of patented and branded prescription devise a pricing strategy? Once patented drugs successfully enter the market, there is little doubt that the producers of these drugs have a considerable degree of monopoly power during the ten-year effective patent life. As shown in Figure 3.1, in the classic monopoly case (i.e., single seller with no substitutes), the profit-maximizing price on the demand curve, D = AR or average revenue, is found at a point where marginal revenue (MR) equals marginal cost (MC). It is the same point that gives profit-maximizing output. The profit-maximizing quantity, *Q*2, is set at point B, where MR equals MC. The profit-maximizing price (point A) is then found by locating the point on the demand curve corresponding to *Q*2. If we assumed that the minimum point on the average cost curve is at B, the shaded area *P*1*P*2AB represents monopolist's above-normal profit. Note that in a perfectly competitive market, where price equals marginal cost equals marginal revenue, long-run competitive equilibrium occurs at point C, where price and average cost are also equal. In that case, there is no above-normal or economic profit in a perfectly competitive industry. As a result, the level of output produced by a perfectly competitive industry (*Q*1) is greater than that produced under monopoly, and the price *P*1 is lower than price *P*2, the price under monopoly.

In the case of near-monopoly (i.e., in the case of weak substitutes), firms have considerable power to set prices well above marginal cost. Much of this price making power is the result of superior product reinforced by strong promotional activities aimed at physicians who make decisions about prescriptions, as well as consumers. In 2005, U.S. pharmaceutical companies spent $29.4 billion on promotion, or 20 percent of domestic sales—$6.8 billion on "detailing" efforts (i.e., face-to-face presentations to physicians) and $18.4 billion on free samples and $4.2 billion on direct-to-consumer advertising (Donohue, Cevasco, and Rosenthal 2007). Scherer (2004) further notes that drug prices significantly above marginal cost have little to do with the firms' efforts to recover the R&D costs. Since past R&D outlays are sunk costs, they are irrelevant for the purpose of current profit-maximizing pricing decisions. However, empirical evidence suggests that real drug prices have an effect on R&D spending. According to a study by Giaccotto, Santerre, and Vernon (2005) that uses industry-level data for the period 1952 to 2001, a 10 percent increase in real drug prices results in a nearly 6 percent increase in pharmaceutical R&D spending.

How do the firms react to potential cut-throat price competition from the generics once the patent expires? Contrary to what one might expect, innovating firms do not respond with preemptive price cuts to discourage the generic firms from entering the market in the first place. Nor do they respond by matching generic price cuts in all segments. Instead, as shown by Vinod and Rao (2000), they maintain monopoly price during the patent period and engage in segment-specific competitive response after the patent expires. And they respond by actually *increasing the price* in the consumer segment, which is less sensitive to price change—in part due to the strength of the brand name built by prior marketing efforts—and more than meet the price cuts in other segments such as hospitals, which are more price sensitive. In effect, they implement a policy of price discrimination, that is, charging different prices to different customers for the same product to maintain above-

Figure 3.1 **Monopoly vs. Competitive Pricing**

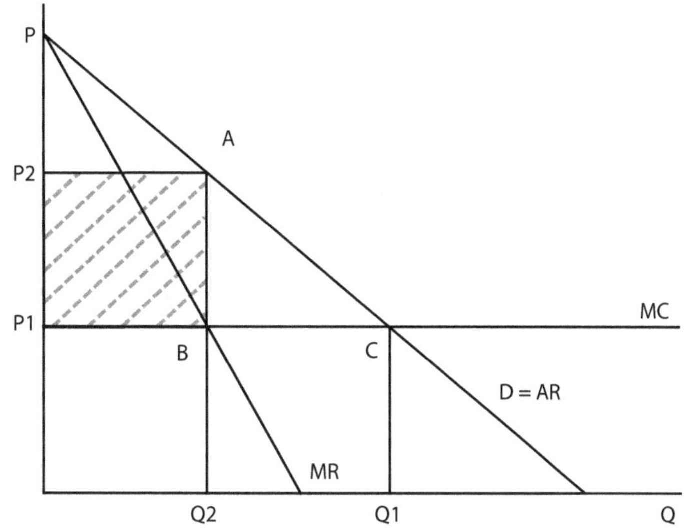

normal profits. Other responses include introducing their own generic version under a different name. More recently, innovating firms have been allegedly paying off the generic producers to stop them from competing with the brand-name prescription drugs. This was the subject of testimony by Jon Leibowitz (2007), Commissioner of Federal Trade Commission, before the Committee on the Judiciary of the U.S. Senate.

THE CASE OF OLIGOPOLY IN THE COMMERCIAL JET AIRCRAFT INDUSTRY

The term oligopoly refers to few sellers. In the case of the commercial jet aircraft industry, there are two firms that dominate the market—Boeing and Airbus. One might refer to this market as a small-number oligopoly or duopoly in which the firms develop, produce, and market differentiated products. Differentiation could take one of many forms—range, capacity, fuel consumption, cockpit design, and the like. The industry has an unusual market structure, in which economies of scale are huge relative to market demand—given the fixed and sunk costs in terms of R&D and marketing. This means minimum average cost is reached at a volume that is a large fraction of the total demand. This is one reason why only two firms survived in this industry. The presence of huge R&D and marketing costs acts as a significant barrier to entry. Given the technological complexity of aircraft production, firms have the opportunity to enjoy the experience curve effects. Another feature of the market structure is that over time Airbus and Boeing began increasing product variety in their offerings in order to serve the needs of different segments. There is no collusive or cooperative behavior by Airbus and Boeing to divide the market among them and achieve monopoly pricing. Each firm appears to act on its own in terms of setting prices and output. In the absence of cooperation, if Airbus and Boeing were producing and marketing undifferentiated planes, one would expect intense price competition or even price wars. Instead, competitive rivalry between the two firms is in R&D, production, and marketing to launch the right plane, for the right segment, at the right price and at the right time.

However, as Douglas and Pavcnik (2001) note, when a firm has closely related products on the market, it is less prone to engage in price competition because of potential cannibalization of its own products. As a result, given the market structure of the industry—huge upfront R&D expenditures and economies of scale—a decision by both firms to serve the same segment could be ruinous to each. As a result, Airbus needs to worry about Boeing's behavior—and vice-versa—in making entry and pricing decisions. This process is referred to as strategic thinking, which is best understood with the tools offered by game theory.

Assume that the entire commercial jet aircraft market is served by only Boeing and Airbus, which constitutes a duopoly. Assume further that each company offers a plane that is similar to its competitor's—Boeing 767 and Airbus 310, for example—and that they face similar cost and demand structures, and each holds, say, 50 percent of the market share. And each firm follows a strategy that is best for itself, so there is no collusion. In this situation, each firm has several alternatives, but their success depends on their rival's reaction. One alternative for Boeing or Airbus is to charge above-normal/monopoly price and hope the rival will follow suit. If it does not and maintains the "normal price"—which is, say, above marginal cost but below monopoly price—it will capture a greater share of the market and increase its profits at the expense of the firm that charges above-normal price. If both charge above-normal price, both will gain. If both charge their normal price, both will earn positive normal profits. The matrix shown in Figure 3.2 describes, hypothetically, for each firm, the profit outcome of alternative strategies depending on how the rival reacts. If both Boeing and Airbus charge normal price, both make normal profits. If Boeing charges above-normal price and Airbus follows, both will make handsome profits. On the other hand, if Boeing charges above-normal price and Airbus stays with the normal price, Boeing stands to lose while Airbus makes above-normal profit. The outcome is reversed when Airbus charges above-normal price, but Boeing stays with the normal price.

Note that both firms can earn above-normal profits (lower right quadrant) if both charge above-normal price by colluding. However, even if we assume collusion, above-normal price is unlikely to be sustainable, given the incentive each firm has to lower the price at the expense of other. So, an important implication of this exercise is, given the incentive each firm has to raise its profits at the expense of the other, normal price and normal profits will prevail. The reader is asked to consult Samuelson and Nordhaus (2010) and Hubbard and O'Brien (2010) for a detailed discussion of duopoly game.

Game Theory

Game theory is also useful in analyzing entry strategies in the context of the commercial jet aircraft market. Esty and Ghemawat (2002) provide a fascinating case study of how Boeing attempted to preempt Airbus A380 in introducing a super jumbo jet, but failed to do so for three main reasons:

1. Given the huge economies of scale relative to the market demand, there is no room in the market for more than one firm.

Figure 3.2 **Airbus 310 Versus Boeing 767: Hypothetical Pricing Strategies**

Airbus 310 Price \ Boeing 767 Price	Normal	Above-normal
Normal	Both Boeing and Airbus make normal profit	Boeing loses money and Airbus makes above-normal profit
Above-normal	Airbus loses money and Boeing makes above-normal profit	Both Boeing and Airbus make above-normal profits

2. Boeing viewed future demand in terms of point-to-point travel, which would solve the problem of congestion at the major airports, whereas Airbus saw growth in terms of hub-to-hub travel between major international airports.
3. Even if Boeing successfully entered into the super jumbo jet market, it would cannibalize the current and future versions of 747 jumbo jets, which had a maximum capacity of over 400. This concern about potential cannibalization not only keeps a firm from engaging in price competition, but also could discourage or delay the introduction of new products.

With respect to the effect of economies of scale on the cost and price, Airbus's A380 super jumbo jet had a projected up-front launch cost of nearly $12 billion—70 percent of Airbus's total revenue and 26 percent of industry revenue in 2000 (Esty and Ghemawat 2002). These are fixed/sunk costs that are the source of the scale economies that need to be recouped eventually. Although the A380's list price was nearly 20 percent higher than the 747's, Airbus has claimed that given the plane's 33 percent greater capacity (555 versus 417 seats in a three-class arrangement) and reduced operating costs, lifetime costs of the plane would be lower.

THE CASE OF MONOPOLISTIC COMPETITION IN THE SOFTWARE MARKET

Monopolistic competition is very common in the United States and other industrialized economies of the world. In monopolistic competition, there are a larger number of firms as compared to oligopoly, albeit fewer than in perfect competition. This means each firm acts independently without having to take into account rivals' reactions, as they would in the oligopoly market structure. There are no entry and exit barriers as might exist in oligopolistic competition. Each firm holds a relatively small share—certainly not a dominating share—of the market, and there is little possibility of collusive behavior. Differentiated products, real or apparent, backed by strong promotional activity (e.g., advertising to build brand name) distinguish monopolistic competition from perfect competition.

Given the above description of the monopolistic market, each firm faces a downward-sloping demand curve (as in monopoly) and it is possible to make monopoly profits in the short run when the technology is new and few firms dominate the market. However, given the ease of entry and relatively low fixed costs, it is also possible that a typical firm would end up making losses. That is, the price would not cover the average cost, a result of too many firms in the market. Eventually, enough firms will exit the market, leaving the survivors to make normal profits. Price equals the average cost over the long run. The software sector, which is populated by many small firms, resembles monopolistic competition. This is evidenced by the fact that the median revenue of the top 25 firms is about $2 billion and the median share of the revenues of these firms is only 1.3 percent. Only three out of the top 25 firms have a share greater than 10 percent. If one were to look at similar data for the top 100 firms, the median revenue would be little over $600 million and the median share would be 0.3 percent (see Table 3.1). Note that

Table 3.1

Global Top 25 Software Companies, 2009 (revenues in millions of U.S. dollars)

	Company	Software Revenues	Share of Top 25
1	Microsoft	$49,453	31.00%
2	IBM	22,089	13.7
3	Oracle	17,560	10.9
4	SAP	11,604	7.2
5	Nintendo	7,245	4.5
6	HP	6,243	3.9
7	Symantec	5,692	3.5
8	Activision Blizzard	4,622	2.9
9	Electronic Arts	4,268	2.7
10	CA	3,936	2.4
11	Adobe	3,361	2.1
12	EMC	3,171	2
13	Konami	2,083	1.3
14	SunGard	2,015	1.3
15	Cisco	1,984	1.2
16	Autodesk	1,982	1.2
17	Dassault	1,605	1
18	BMC	1,585	1
19	Namco Bandai Games	1,569	1
20	Sage	1,496	0.9
21	Ubisoft	1,486	0.9
22	Take 2 Interactive	1,452	0.9
23	NCR	1,431	0.9
24	Sega Sammy Holdings	1,407	0.9
25	Intuit	1,383	0.9
	Total	$160,722	

Source: Balder Verbene, "Global Software Top 100—Edition 2009," Software Top 100, November 10, 2009, http://www.softwaretop100.org/global-software-top-100-edition-2009; and authors' calculations.

the personal computer segment of the computer industry has also evolved from domination by a few oligopolistic firms like Apple and Compaq to one with many firms, each with a relatively small share of the market and no monopoly profits.

APPLE'S IPHONE PRICING STRATEGY

Apple's iPhone pricing strategy suggests that it follows *skimming* and *versioning* (Sliwinska, Ranasinghe, and Kardava 2007–2008). Apple launched its iPhone in the cellular phone market initially dominated by Nokia, Motorola, Sony Ericsson, and Samsung. Apple entered the market with a game-changing innovation in the early maturity stage, at least in the industrialized countries like the U.S. The iPhone combined smart phone capabilities with touch technology, wireless Internet communications—e-mail, search engines, Web sites—computer, camera, and PDA.

Apple introduced two iPhone models—4GB at $499 and 8GB at $599. Most of the iPhone users are relatively technology savvy and can afford to spend money on smart phones. Apple followed the skimming price strategy, but with a difference from the traditional way of skimming, which is to gradually reduce the price over PLC before extending the PLC with an improved version. Apple discontinued the $499 model and reduced the price of the premium model from $599 to $399. Moreover, every new generation model of the iPhone was introduced at either the same price as its predecessor or at a lower price—$399 (2002), $299 (2003), $299 (2004), and $249 (2005). The idea behind this strategy is to maintain sustained growth with improved and attractively packaged versions of the same product. Apparently the additional revenue net of cannibalization of older models from each new version far exceeds the additional cost of producing and marketing them. So rapid is Apple's introduction of newer versions, competitors have little time to respond, especially in light of Apple's formidable brand name and its many versions aimed at different segments of the market.

PRICE DISCRIMINATION

Price discrimination involves charging different prices to different customers for the same product when the cost of supplying different consumers is the same. Price discrimination also occurs when the cost of supplying a product to different customers varies, but the firm charges the same price. Price discrimination based on what consumers are willing to pay eventually results in greater profits than if the firm charged a single price to all customers.

DEGREES OF PRICE DISCRIMINATION

A perfectly discriminating monopolist charges a different price to each customer in order to maximize profits. This is known as first-degree price discrimination. However, this is not generally feasible. Second- and third-degree price discrimination are more common. Second-degree price discrimination refers to a case where every potential customer chooses from a menu of related products. This occurs when products like smart phones are sold in product lines or when products are sold in different versions, such as books that are available in hardcover or paperback. Third-degree price discrimination occurs when marketers charge different prices to different *groups (segments)* of customers with different price elasticities of demand. Successful price discrimination does not allow arbitrage, that is, customers who pay a lower price for the same product cannot turn around and sell it in the higher-price market to make a profit. Assuming that the marginal cost of producing a product is same across customer groups, profit-maximizing quantities are determined such that marginal revenue for each group is equal to marginal cost and profit-maximizing prices for those quantities are found on the demand curve.

EXPLOITING SEGMENT-SPECIFIC DIFFERENCES IN ELASTICITY

Figures 3.3a and b illustrate the commonly practiced third-degree price discrimination. The two segments in a market differ by the degree of price elasticity. The steeper demand curve (D_A) in Figure 3.3a represents a market segment that is relatively inelastic or less price sensitive. Marginal cost (MC) is a horizontal line, meaning MC is constant. We find the profit-maximizing quantity for each segment where the MC for that segment intersects the marginal revenue (MR) for that segment. So, the profit-maximizing quantity for the inelastic segment is Q_A where MC intersects with MR_A. Similarly, the profit-maximizing quantity for the elastic segment as shown in Figure 3.3.b is Q_B where MC intersects with MR_B. To determine the profit-maximizing prices, look at the respective demand curves of each segment. P_A and P_B are the profit-maximizing prices for the two segments, respectively. Since average cost is the same for both of the segments, the shaded area represents above-normal profits for the discriminating monopolist in each segment. The point is that the monopolist's profits in the two segments combined are greater than if he charged a single price.

Price discrimination is pervasive in the marketing of products and services. Theater tickets sold at significant discounts to the elderly, deep discounts by airlines, resort hotels, and off-peak rates by power companies are just a few examples of price discrimination.

DISGUISED PRICE DISCRIMINATION

Marketers also practice what may be referred to as *disguised* price discrimination. This takes the form of a pharmaceutical company selling branded prescription drugs at a significantly higher price in the United States than the same drug produced by the same firm and sold at a significantly lower price—under a different name and package, for example—in a price-sensitive developing-country market. Disguised price discrimination may entail additional costs involving change of color, packaging, and other features. However, it pays for the firm to engage in discrimination as long as additional revenues derived from the price discrimination exceed additional costs of price discrimination.

Under certain conditions, marketers mimic price discrimination over product life cycle (PLC). Assume that a firm has developed a new and novel product with a strong brand

Figure 3.3 **Profit-Maximizing Price Discrimination**

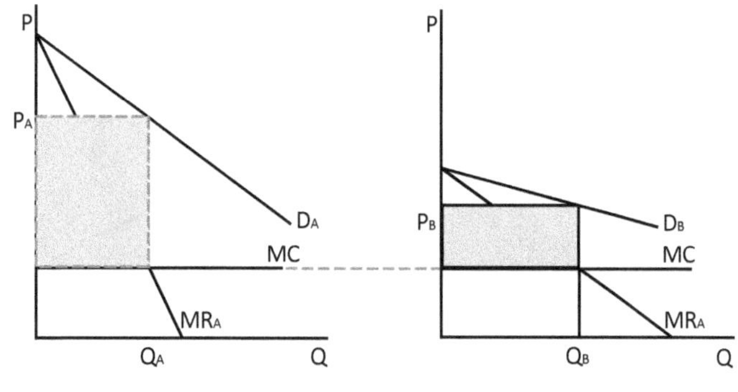

(a) Relatively inelastic market segment (b) Relatively elastic market segment

name. The firm holds a highly enforceable twenty-year patent and there is no potential competition as far as the eye can see. In effect, it is a twenty-year monopoly facing a strong appropriability regime (see Figure 1.2). Under these conditions, the firm, as shown in Figure 3.4 (a and b), has an incentive to mimic price discrimination by reducing the price gradually over the life of the product. All one needs to do is to think of the usual four PLC stages—introduction, growth, maturity, and decline—as representing four segments of consumers with progressively increasing price sensitivity.

PRICING INFORMATION

Information markets have certain special characteristics that present significant challenges to developing profitable pricing strategies (Shapiro and Varian 1999; Varian, Farrell, and Shapiro 2004; and Miller 2011). First, information is expensive to produce and cheap to reproduce at constant per unit cost. That is, the marginal cost of making and distributing additional copies can be extremely low or near-zero and constant. Second, fixed costs of production and promotion associated with the first copy are sunk costs and therefore irrelevant to pricing. Third, because of large upfront fixed cost associated with the production of the first copy and very low or near zero marginal cost of subsequent copies, there are significant short-run economies of scale. While many high-tech products share this feature, information products represent an extreme version of it. Fourth, there is generally no capacity constraint for reproducing additional copies. Fifth, given that information products can be easily shared by competitors and consumers alike at little or no cost—despite copyright and patent protection—firms in this market generally face weak appropriability regime. As a result, competition drives prices down to marginal cost—which is near zero—and the pioneer firm is unable to recover its investment, much less make a profit. This is illustrated in Figure 3.5.

Assume that an entrepreneur develops an antivirus applications software program that is particularly useful. He incurs $60,000 in fixed costs, the cost of producing the first CD. The marginal cost of producing a CD is a constant $1. If he were operating in a perfectly competitive market the price would be $1, which is equal to the marginal

Figure 3.4 **Price Discrimination over Product Life Cycle (PLC)**

(a) Product Life Cycle

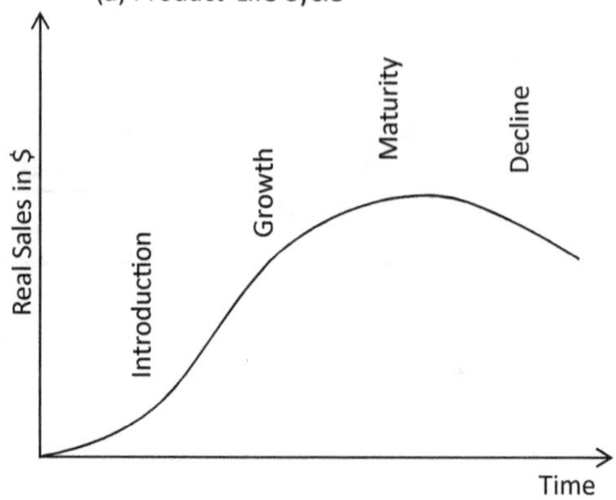

(b) Pricing over Product Life Cycle

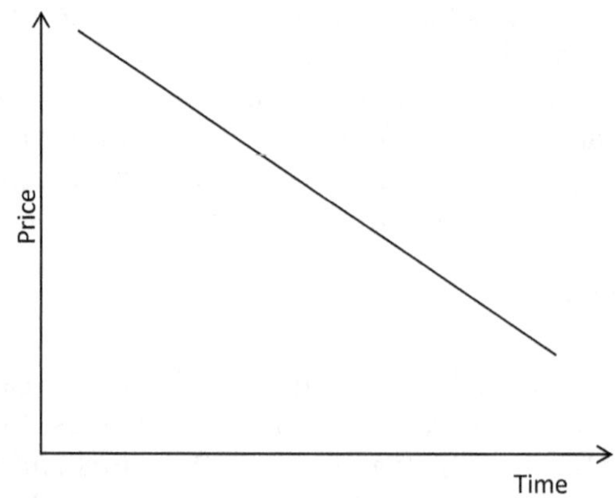

cost (MC). At this price our entrepreneur sells 12,000 units and generates $12,000 revenue. However, the total cost of producing 12,000 units is $72,000 (marginal cost of $1 times 12,000 plus $60,000 fixed cost = $72,000), which results in $60,000 economic loss shown by the shaded area in Figure 3.5. So, perfectly competitive markets do not work well, if they work at all, in the case of information goods. Monopolistic competition prevails. As discussed earlier, many firms enter and exit in this market structure over time. Ultimately, the price surviving firms charge should cover the average cost. At $13 per CD our entrepreneur sells 5,000 units and generates $65,000 revenue, which covers his fixed cost of $60,000 and variable cost of $5,000. Note that he makes his normal return, but his economic profit is zero.

BUNDLING

Bundling occurs when two or more different products are sold together for a single price. Information goods make especially good candidates for bundling because the marginal cost of adding an extra good to a bundle is extremely low, or near zero.

Figure 3.5 **Pricing of Information Under Monopolistic Competition**

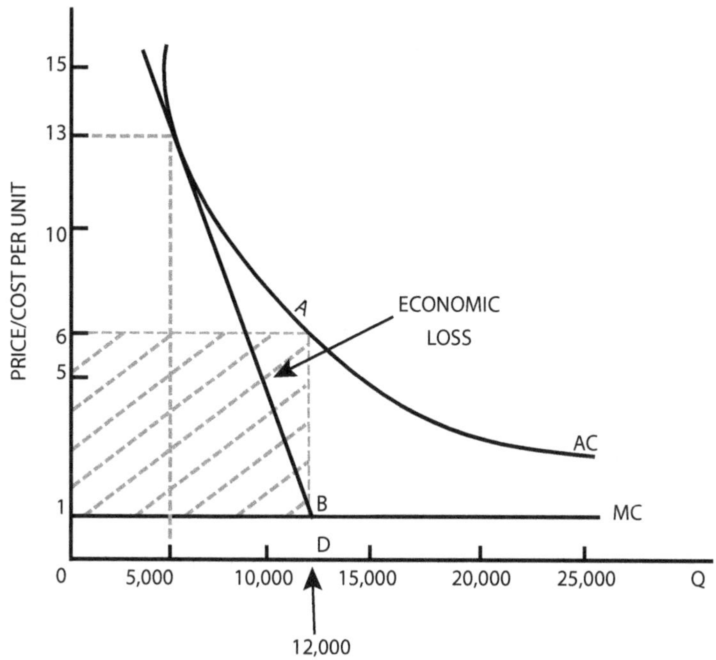

Bundling works because it reduces the dispersion of willingness to pay and makes the demand curve more flat or more price sensitive. Consider this example: Consumers A and B value the paper and online versions of the *New York Times* differently. Consumer A, who enjoys the paper version of the *Times*, is willing to pay $200 for a year's subscription, but only $100 for the online version of the same, whereas consumer B, who finds it more convenient to read online, is willing to pay $200 for the online version and only $100 for the paper version. At $200, consumer A will subscribe to the paper version only and consumer B will subscribe to the online version only. So, if each version is sold separately, the *Times* will earn $300. On the other hand, if the *Times* bundles the two versions at a price of $200—assuming that the willingness to pay for the bundle is the sum of the willingness to pay for the two versions separately—it will make $400 each. Note that this is another form of disguised price discrimination. If the *Times* somehow were able to implement perfect price discrimination, it would need to develop two different prices for A and B.

EXPERIENCE CURVE PRICING

As discussed in chapter 2, experience curve effects refer to systematic reductions in unit cost by a fixed percentage every time production doubles. The two sources of experience curve effects are: (1) cost savings that come from learning by doing, and (2) economies of scale. Both of these sources are particularly prominent in the production and marketing of high-tech products and services. There is greater opportunity to exploit learning curve effects when a technologically complex task is repeated. It is important to note that cost reductions attributed to learning by doing come from many other sources other than learning by labor in the plant. They include improvements over time in process technology, product design, marketing knowledge, and economies of joint production and marketing or economies of scope. And high-tech products typically involve large upfront fixed/sunk costs associated with R&D, setting up production facilities, and promotion, which are the sources of economies of scale.

As noted earlier, the Airbus A380's developmental costs alone were in the range of $12 billion. Similarly, producers of prescription drugs spend $800 million to bring a new drug to the market, which has a chance of success of one in ten. In certain industries, like the auto industry, minimum optimal size of the plant—about $250 million—imposes greater upfront fixed costs than is normally the case in manufacturing.

A firm that recognizes the benefits of the experience curve effect and chooses to exploit it gains competitive advantage not only in terms of lower unit costs, but also by building marketing assets in the form of distribution network, brand name, and relationships with customers and suppliers—all serving as barriers to competition. The presence of significant switching costs and network externalities, which are common in the information and communication technology (ICT) sector, would further reinforce the firm's first-mover advantage. From the standpoint of pricing strategy, the experience curve effect—when it is significant and predictable with a high level of confidence—provides the first-to-enter firm considerable flexibility in pricing over the PLC. If the firm believes that it faces a strong appropriability regime, it is likely to set its initial price higher than can be justified by its estimated experience curve. That is, it could set its initial price as if the slope of the experience curve were flatter than the slope of its estimated curve. However, while such a strategy produces bigger profits in the short run, it almost always entails the risk of inviting competition. This is what happened to Kodak in the early 1950s when its high initial price of film invited Fuji and Agfa into the market (Bogue III and Buffa 1986; Czepiel 1992).

PENETRATION PRICING

Alternatively, if a firm believes that it faces a relatively weak appropriability regime, it would deter the threat of new entrants by setting its initial price lower than indicated by its estimated curve. This strategy is called "penetration pricing." The threat of new entrants, despite their cost disadvantage over the incumbent firm, cannot be exaggerated and pricing strategy alone cannot deter entry. This is because new entrants could overcome their cost disadvantage with superior product design and process technology that enables them to produce a better product at lower cost. The new entrants, as shown in Figure 3.6, can shift the entire curve downward. Also, recall from chapter 2 that most of the cost reductions from learning-by-doing occur in the early stages of production. Thus, the incumbent firms that fail to maintain their technological lead will quickly find themselves at a competitive disadvantage not only in terms of cost, but also in terms of product innovation.

Figure 3.6 **New Entrants Have the Advantage of Newer Technology**

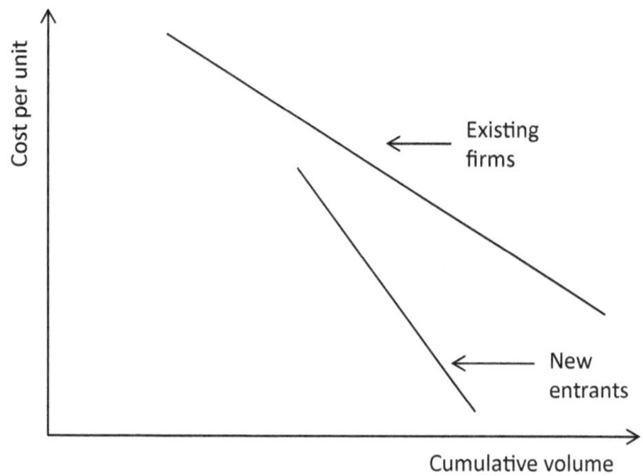

BANDWAGON EFFECTS AND PRICING

Jeffrey H. Rohlfs (2003), in his seminal work *Bandwagon Effects*, provided the theoretical framework for understanding the implications of bandwagon demand and bandwagon supply for pricing of many high-tech products and services, ranging from fax machines to CDs to PCs to the Internet. On the demand side, bandwagon effects occur when a product or service is worth more to each consumer as the user set expands. These effects are known as "network externalities," or external demand-side scale economies. Telephone, e-mail, and personal computers are examples of products and services subject to bandwagon effects. Another category of bandwagon effects—known as complementary bandwagon effects—occur when consumers of a product benefit from greater availability of complementary products, like consumers of personal computers who benefit from increases in the availability of software.

In the simplest case, Rohlfs's theory of bandwagon demand suggests that the demand curve for bandwagon products and services—unlike the normal inverse demand curve for nonbandwagon products and services—is an inverted U, as shown in Figure 3.7.

This is because as the user set expands, benefits increase the value of service to each user and the reservation price (i.e., the highest price a consumer is willing to pay) of the marginal user *increases* until a critical mass is reached. Thus, the left part of the curve in Figure 3.7 is upward sloping. Note that the downward sloping part of the figure suggests that additional users value the service progressively less, as is the case with nonbandwagon services.

On the supply side, because of the need to reach a critical mass of demand quickly in order to benefit from the first-mover advantages, the supplier of service has to be concerned with the start-up problem of getting the initial user set to sign up. Even in the case of monopoly—pure competition seldom exists in high-tech markets—it may be a better start-up strategy to sacrifice short-run profits for long-term gain and set the price low enough to attract the initial user set. Consumers stand to benefit in the short run as well as in the long term—from lower prices in the short run and

Figure 3.7 **Illustrative Demand Curve For Bandwagon Services**

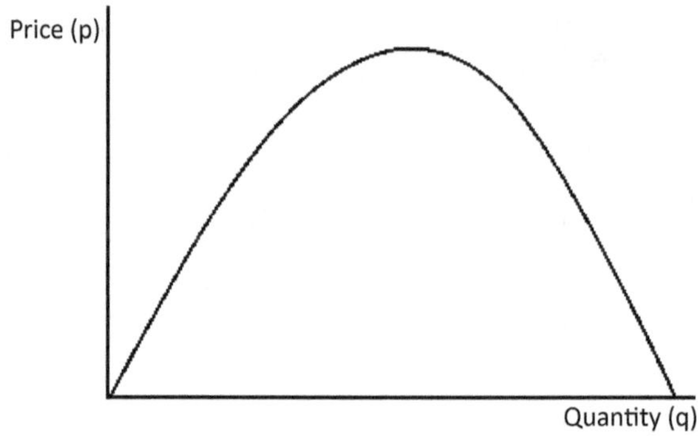

Source: Rohlfs (2001).

from bandwagon effects in the long run. The strategy of low initial price to solve the start-up problem becomes central in the oligopolistic market structure where rivalry to be the first to take advantage of bandwagon effects is intense.

The literature is replete with examples of the critical role of pricing in the bandwagon race. In the mid-1980s, Apple, in its attempt to benefit from expanded use of Apple computers, accepted significant losses by keeping prices low. However, this otherwise defensible strategy was largely undone by its choice of excessive vertical integration—which reduced flexibility to users—and its refusal to license the operating system (Rohlfs 2003). Apple's strategy fell short on both counts—the former prevented Apple from achieving dominance in the hardware market and the latter resulted in Windows, not Macintosh, becoming the industry standard. Note that, despite its virtual monopoly of the Windows operating system, historically Microsoft priced its successive versions of Windows relatively low to exploit the bandwagon effects and maintain its dominance in applications software. Gilbert and Katz (2001) cite Schmalensee's estimate (1999) suggesting that the monopoly price of Windows should have been sixteen times the price actually charged.

Pricing also played a significant role in the success of Matsushita's VCRs based on the VHS standard in the bandwagon race against Sony's VCRs based on the Beta standard. Matsushita recruited RCA to market its VCRs in the United States, and Sony allowed its VCRs to be sold under Zenith's brand name, one of the largest names in the manufacturing of televisions. While playing time and tape compactness may have proved to be the decisive factors in the ultimate success of Matsushita's VCRs, aggressive pricing and licensing strategies following their introduction in 1977 played a significant role in exploiting the bandwagon effects (Liebowitz and Margolis 2002).

MARKET FOR TECHNOLOGIES

Several authors, such as Baumol (2002) and Arora, Fosfuri, and Gambardella (2001), have written about the emerging "market for technologies," as opposed to the market for products and services. How does one describe the market structure for technologies? How are prices determined in such a market structure? What is the nature of the appropriability regime in the market for technologies? These are but a few of the difficult questions that confront firms engaged in the production and sale of technologies.

PRICING OF TECHNOLOGY

If pricing of high-tech products and services is challenging, pricing of technology is doubly challenging. The "market for technology," when it exists, cannot easily be analyzed in terms of any of the market structures examined earlier. Given the problem of moral hazard, transactions are few and extremely confidential. As a result, prices are determined by negotiation.

Intellectual Property

Teece (2000) suggests that in any valuation of technology, distinction must be made between the technology or innovation and the IP that incorporates that technology. IP is a legal right in the form of patents and copyrights, which may not cover all of the technology, further complicating the pricing problem. Moreover, when there is ambiguity with respect to IPRs, there is likely to be, ceteris paribus, greater spread between the bidding price and the asking price than would be the case otherwise.

Complementary Manufacturing

Still another factor that complicates pricing of technology is the degree to which complementary manufacturing and marketing assets are important for the commercial success of the technology. Returning to the example noted earlier, biotech firms—whose sole activity was to produce innovations—have to split the gains with the traditional pharmaceutical firms that possess complementary production and marketing assets in the form of distribution and brand name. Thus, if the buyer of the innovation holds significant complementary assets on which commercial success depends, the seller of the innovation has to split the expected gains with the buyer appropriately. On the other hand, if complementary assets are readily available on the market, the seller would have greater bargaining power.

Licensing

Licensing is the most common form of marketing technology. In general, a strong appropriability regime is conducive to greater control over the terms of the license as well as protection of technology. If, on the other hand, appropriability is weak, the firm may pursue outright sale of technology or turn the technology into a differentiated product or service through marketing effort.

Typically, licensing involves many restrictions on the buyer as well as seller. The seller (licensor) could impose restrictions on the buyer (licensee) with respect to exclusivity, duration of license, geographic coverage, cross-licensing, transfer of technologies to third parties, and the like. Similarly, the seller is required to cooperate, on a continuing basis, in the transfer of technology that could take several years, depending on the complexity and the degree to which it is codified.

Teece (2000) summarizes four broad licensing objectives:

1. *Efficient commercialization:* Firms with technologies, but lacking complementary marketing assets in the form of a sales force, distribution network, or brand name, find out-licensing attractive. Similarly, firms with complementary marketing assets but lacking the technology find in-licensing an attractive option. An example of this is the out-licensing of technology to traditional pharmaceutical firms with vast production and marketing assets by biotechnology firms in their early days.
2. *Technology exchange:* Technology exchange occurs through cross-licensing within a field of use. Exchange of complementary technologies without running into infringement problems is a strong motive behind cross-licensing. In general, components of technology are priced separately and netted out to determine cash payments, if any.
3. *Strategic licensing:* Licensing can be used strategically to establish a new industry product standard to take advantage of the bandwagon effects (the early decision by Microsoft to license the Windows operating system, for example); to gain access to new international markets; and to induce suppliers to adapt the new technology (e.g., licensing of technology for CDs by Philips).
4. *Royalty generation:* The motive here is simply the generation of cash by licensing technologies that have been on the shelf but for which the firm has no current or future use for them. Firms also license the technologies that are in the mature stage of the PLC in the home market, but are valuable to firm in, say, developing country markets where the product might be in the introductory stage.

A variety of licensing and cross-licensing arrangements are possible, depending on the nature of innovation, the industry, the importance of complementary assets, and the like. Contracts can be exclusive or nonexclusive, although the latter are more common. While an exclusive contract demands a higher price than a nonexclusive contract, the seller loses the flexibility and the ability to exploit unanticipated opportunities. The payment could be an up-front lump sum or royalty on the sales, or a combination of the two. If the licensor's objective is to generate cash in the short run, if he sees no particular use for the technology, and is uncertain about its commercial success, he is likely to favor up-front lump-sum payment. On the other hand, if he expects the technology to be a significant commercial success, other things being equal, he is likely to favor royalty on sales. Note that the royalty option does impose monitoring costs on the parties concerning the sales base, for example.

Quantifying the value of the license requires detailed information on several aspects of the technology. If the technology in question has to do with process innovation, estimates of the present value of net production and capital cost savings are required. Capital cost savings are those savings attributable to the new process relative to the savings that would have occurred if the new process had not been developed. On the other hand, if the technology is a product- or service-related innovation, one needs to estimate the present value of the expected future sales attributable to the innovation, net of transfer and transaction costs for the relevant period, geographic area, and the like.

Teece (2000) provides a number of detailed case studies of licensing and cross-licensing in the computer, semiconductor, telecommunications, and glass industries. He notes that the main objective of cross-licensing by firms like IBM, Texas Instruments (TI), and Hewlett Packard is to obtain "design freedom." That is, given the competitive pressures to speed up innovation and their inability to do everything in-house, firms like IBM need the rights to use technology created by others. Cross-licensing agreements at TI typically cover groups of patents with designated "fields of use," including existing and new patents developed within a fixed term. This is a strategic motive behind cross-licensing. Where royalty payments are concerned, at TI they are netted out according to what is known as a "proud list" procedure. Each firm produces a sample list of its most valuable patents, which serves as the basis to value the entire group of patents. The value of each patent in the sample is evaluated with respect to expected sales derived from it, as well as such qualitative factors as legal validity and enforceability of the patent and technological significance. The dollar amount of the value of all sample patents so calculated is then expressed as a percentage of each licensee's total sales—the royalty rate—and applied to the licensee's sales for the term of the license.

What is the "right" price or license fee from the point of view of the innovator-firm holding the patent? Baumol (2002) asked this question and provided a pricing principle—the efficient component-pricing rule, or parity principle—that addresses the issue of profit maximizing price as well as efficiency implications. Baumol applies this principle to the case of a patent holder who is also the supplier of some or all of the final products that use the innovation. According to the parity principle, the price or the license fee a patent holder charges to a final product customer is equal to the final product price charged by the patent holder (who also supplies the final product) minus its incremental cost of remaining inputs. Alternatively, the license fee must equal the direct incremental cost to the patent holder resulting from the use of the innovation by others plus the associated incremental opportunity cost or the loss of profit resulting from licensing to the competitors. Simply put, license price of innovation, following the basic economic principle, should equal incremental cost including the incremental opportunity cost. Note that the competitors' price of the final product sales would have to be higher than the patent holder's final product price by the amount of license fee the competing firms would have to pay. This, combined with the sale of licenses, contributes to the recovery of sunk costs associated with innovation. Baumol considers several variants of the parity principle depending on the particular market structure faced by the patent holder and/or the competing suppliers.

SUMMARY

Producers of high-tech products and services operate in imperfectly competitive structure. That is, products and services are highly differentiated by means of R&D

and complementary promotion. Differentiation provides firms with a degree of market power that in turn allows them to become price setters rather than price takers. Pricing strategies of firms vary with the particular form of imperfect competition. Economists generally distinguish between three types of imperfectly competitive markets: (i) monopoly (single producer with no real substitutes, at least in the short run like patented drugs); (ii) oligopoly (few producers with varying degrees of product differentiation such as autos); and (iii) monopolistic competition (many producers with real or apparent differentiation like personal computers). Within these three market structures, pricing strategies vary depending on the intensity of rivalry and the threat of *potential* competition (the case of pharmaceuticals); opportunity to achieve first-mover advantages by exploiting scale and experience curve effects (the case of the commercial jet aircraft industry); skimming and versioning (the case of Apple's iPhone); and disguised price discrimination (the case of pharmaceutical products that carry higher prices in the developed-country markets and lower prices in the developing-country markets). Pricing information presents particularly difficult challenges because of the many unique characteristics of information markets—high costs of production and near-zero marginal cost of reproduction, ability to share information by consumers and competitors alike at little or no cost, and so on. Versioning, bundling strategies, and disguised price discrimination become especially important in information markets. Finally, when firms choose to market "technologies" rather than products and services in the usual sense, they are operating in a technology market that cannot be examined in terms of any of the market structures noted above. In technology markets, prices are negotiated in the form of licensing, cross-licensing, or technology exchange.

KEY TERMS

bandwagon effects
bundling
cross-licensing
disguised price discrimination
efficient component-pricing rule
experience curve pricing
information goods
monopolistic competition
monopoly
oligopoly
parity principle
price discrimination
strategic licensing
sunk costs
technology market

DISCUSSION QUESTIONS

1. Explain how pricing strategies for high-tech products and services are guided by features that distinguish them from commodity-like products.
2. Discuss the essential features of the three imperfectly competitive market structures—monopoly, oligopoly, and monopolistic competition—and their respective application to: (i) the prescription drug segment of the pharmaceutical industry, (ii) the commercial jet aircraft industry, and (iii) the software industry.

3. (a) Explain the concept of price discrimination including first-degree, second-degree, and third-degree price discrimination.
   (b) What are the key conditions for successful price discrimination?
4. (a) What is meant by disguised price discrimination?
   (b) How do marketers mimic price discrimination over the product life cycle?
5. (a) What are the special characteristics of information markets?
   (b) What is bundling and why do information goods make good candidates for bundling?
6. Explain why the experience curve effect, when it is significant and predictable, provides flexibility in pricing over the product life cycle.
7. (a) What is meant by the "market for technologies" and what is the nature of its appropriability regime?
   (b) Why is licensing the most common form of marketing technology? And what are the key objectives in licensing?
8. Explain why pricing of technology is often done by negotiation.

REFERENCES

Arora, Ashish, Andrea Fosfuri, and Alfonso Gambardella. 2001. *Markets for technology: The economics of innovation and corporate strategy*. Cambridge, MA: MIT Press.

Baumol, William J. 2002. *The free market innovation machine*. Princeton, NJ: Princeton University Press.

Bogue III, Marcus C., and Buffa, Elwood. 1986. *Corporate strategic analysis*. New York: Free Press.

Czepiel, John A. 1992. *Competitive marketing strategy.* Englewood Cliffs, NJ: Prentice Hall.

Donohue, J.M., M. Cevasco, and M.B. Rosenthal. 2007. A decade of direct-to-consumer advertising of prescription drugs. *New England Journal of Medicine* 357: 673–681.

Douglas A. Irwin, and Nina Pavcnik. 2001. Airbus versus Boeing revisited: International competition in the aircraft market, NBER Working Papers 8648, National Bureau of Economic Research, Inc.

Esty, Benjamin C., and Pankaj Ghemawat. 2002. Airbus vs. Boeing in superjumbos: A case of failed preemption (February). HBS Strategy Unit Working Paper No. 02-061; HBS Finance Working Paper No. 02-061.

Giaccotto, C., R.E. Santerre, and J. A. Vernon. 2005. Pharmaceutical pricing and R&D growth rates. *Journal of Law and Economics* 48 (1): 195–214.

Gilbert, Richard J., and Michael L. Katz. 2001. An economist's guide to U.S. v. Microsoft. *Journal of Economic Perspectives* 15 (2): 25–44.

Hubbard, R. Glenn, and Anthony Patrick O'Brien. 2010. *Microeconomics*. 3d ed. Upper Saddle River, NJ: Prentice Hall.

Leibowitz, Joe. 2007. Prepared statement of the federal trade commission before the committee on the judiciary of the United States Senate on anticompetitive patent settlements in the pharmaceutical industry: The benefits of a legislative solution (January 17). http://www.judiciary.senate.gov/hearings/testimony.cfm?id=e655f9e2809e5476862f735da11e5c9b&wit_id=e655f9e2809e5476862f735da11e5c9b-1-1.

Liebowitz, Stan J., and Stephen E. Margolis. 2002. Beta, Macintosh, and other fabulous tales. In *Famous fables of economics: Myths of market failures*, ed. Daniel F. Spulber, pp. 110–121. Malden, MA: Blackwell.

Miller, R.L. 2011. *Economics Today.* 15th ed. Boston: Addison-Wesley, 680–682.

Rohlfs, J.H. 2003. *Bandwagon effects in high-technology industries*. Cambridge, MA: MIT Press.

Samuelson, P.A., and W.D. Nordhaus. 2010. *Economics*. 19th ed. New York: McGraw-Hill, 195–199.

Scherer, F.M. 2004. The pharmaceutical industry—prices and progress. *The New England Journal of Medicine* (August 26), 351: 927–932.

Schmalensee, Richard L. 1999. Schmalensee testimony, available at http://www.microsoft.com/presspass/trial/schmal/schmal.asp.

Shapiro, C., and H.R. Varian. 1999. *Information rules: A strategic guide to the network economy.* Boston: Harvard Business School Press.

Sliwinska, Dorota, Jani Ranasinghe, and Inga Kardave. 2008. http://aeunike.lecture.ub.ac.id/files/2012/04/case-Kel.13.pdf (accessed January 28, 2013).

Teece, David. 2000. *Managing intellectual capital.* Oxford: Oxford University Press.

Varian, H.R., R. Farrell, and C. Shapiro. 2004. *The economics of information technology: An introduction.* Cambridge, UK: Cambridge University Press.

Vinod, H.D., and P.M. Rao. 2000. R&D and promotion in pharmaceuticals: A conceptual framework and empirical exploration. *Journal of Marketing: Theory and Practice*, (8) 4: 10–20.

4 IP Primer

A Legal Perspective

Chapter Overview

Information or knowledge embodied in a tangible object such as a book, software code, or a machine constitutes intellectual property, which may be protected by law. As we defined intellectual property in chapter 1, we are using the term to refer more specifically to all technology-based, intangible assets of a firm—an idea or a design for a new product or process, the writing of software code, or a company's logo, trade name, and the like, which may be protected as a property right created by statute, case law, or a combination of both. This chapter provides a background on the various tools used to protect intellectual property—specifically copyrights, patents, and trademarks—and then analyzes how the views on these tools have changed, demonstrated by the open source movement and various public policy reforms.

Legally Enforcing Intellectual Property Rights

Purposes of Intellectual Property Laws

Laws that protect intellectual property serve two primary purposes. The first is to grant legally recognizable rights to creators and other producers of the knowledge and information embodied in goods and services to control their use and distribution. The second is to promote, as a matter of public policy, innovation and the dissemination and application of its results in order to stimulate technological change and the enhanced productivity that it produces.

Available Remedies

In cases of infringement of an owner's intellectual property rights, there are a number of remedies available to legally enforce these rights. Some examples are given below in the discussion of specific types of intellectual property, but in general the remedies fall into five general categories:

1. Court orders to stop or prevent a violation, known as injunctions;
2. Monetary damages;
3. Criminal sanctions, including fines and possible imprisonment in cases of willful infringement;
4. Destruction of infringing products;
5. Border measures to stop suspected imported infringing goods from entering or being distributed within a country.

Types of Intellectual Property

The Convention Establishing the World Intellectual Property Organization (WIPO), concluded in Stockholm on July 14, 1967 (Article 2 (viii)), provides that intellectual property shall include rights relating to:

- "Literary, artistic and other works
- Performances of performing artists, phonograms, and broadcasts
- Inventions in all fields of human endeavor
- Scientific discoveries
- Industrial designs
- Trademarks, service marks, and commercial names and designations
- Protection against unfair competition, and all other rights resulting from intellectual activity in the industrial, scientific, literary, or artistic fields" (WIPO Industrial Property, 3)

We shall focus on a subset that encompasses copyrights, patents, trademarks, and trade secrets.

Table 4.1 provides a summary of major international treaties and their provisions with respect to copyrights and patents. The three main features of the Agreement on Trade-Related Aspects of Intellectual Property Rights (TRIPs) are described in Exhibit 4.1.

Copyright Branch of Intellectual Property

The areas of literary, artistic, and scientific works belong to the copyright branch of intellectual property. Although software is not specifically listed among the examples of such copyrightable works, software programs and related documentation customarily receive copyright protection under the copyright laws of a number of countries.

Table 4.1

Major International Intellectual Property Treaties

Protection Type	Major International Treaties	Description
Copyrights	1. Berne Convention for the Protection of Literary and Artistic Works, 828 UNTS 221	1. Signatories must recognize the copyrights from their own country and from other signatory countries in the same way
	2. WIPO Copyright Treaty 36 ILM 65	2. Extends copyright protection under the Berne Convention to computer programs and compilations of data, among other things
Patents and Trademarks	1. Paris Convention for the Protection of Industrial Property 21 UST 1583; 24 UST 2140; TIAS 6923, 7727; 6 ILM 806	1. The first major international treaty designed to help the people of one country obtain protection in other countries for their intellectual creations
	2. Madrid Agreement concerning International Registration Marks 828 UNTS 390	2. Makes it possible to protect a mark in a large number of countries by obtaining an international registration
	3. Patent Law Treaty 39 ILM 1047	3. Harmonizes and streamlines formal procedures with respect to national and regional patent applications and patents
	4. Patent Cooperation Treaty 28 UST 7645; TIAS 8733	4. Makes it possible to seek patent protection for an invention simultaneously in each of a large number of countries, utilizing an international application as a preliminary step before making national filings—however, there is no international patent
	5. European Patent Convention 13 ILM 268	5. Provides for the granting of European patents via a harmonized procedure

Source: World Intellectual Property Organization, "WIPO-Administered Treaties," http://www.wipo.int/treaties/en/index.jsp.

As WIPO explains: "Unlike protection of inventions, copyright law protects only the *form of expression* of ideas, not the ideas themselves. The creativity protected by copyright law is creativity in the choice and arrangement of words, musical notes, colors and shapes. So copyright law protects the owner of property rights against those who copy or otherwise take and use the form in which the original work was expressed by the author" (emphasis in the original) (WIPO Copyright 5).

FEDERAL PROTECTION

In the United States, copyrights are recognized in the Constitution (Article 1, Section 8, Clause 8) and by federal statutory law under Title 17 of the United States Code. The purpose of copyright protection for the particular expression of an idea is to prevent an unauthorized party from reproducing, distributing, publicly displaying, or performing the copyrighted work or creating a derivative work based on the copy-

Exhibit 4.1 **TRIPs Agreement**

"The Agreement on Trade-Related Aspects of Intellectual Property Rights 33 ILM 81 (World Trade Organization) establishes minimum levels of protection that each government has to give to the intellectual property of fellow WTO members.

The TRIPS Agreement, which came into effect on January 1, 1995, is to date the most comprehensive multilateral agreement on intellectual property.

The areas of intellectual property that it covers are: copyright and related rights; trademarks including service marks; geographical indications including appellations of origin; industrial designs; patents including the protection of new varieties of plants; the layout-designs of integrated circuits; and undisclosed information including trade secrets and test data.

In addition the TRIPS Agreement provides for certain basic principles, such as national and most-favored-nation treatment, and some general rules to ensure that procedural difficulties in acquiring or maintaining intellectual property rights do not nullify the substantive benefits that should flow from the TRIPS Agreement. The obligations under the TRIPS Agreement will apply equally to all Member countries, but developing countries will have a longer period to phase them in."

The three main features of the Agreement are:

1. *Standards.* The TRIPS Agreement sets out the minimum standards of protection to be provided by each member country. Each of the main elements of protection is defined, namely the subject-matter to be protected, the rights to be conferred and permissible exceptions to those rights, and the minimum duration of protection. The TRIPS Agreement sets these standards by requiring, first, that the substantive obligations of the main conventions of the WIPO, the Paris Convention for the Protection of Industrial and the Berne Convention for the Protection of Literary and Artistic Works in their most recent versions, must be complied with. Secondly, the TRIPS Agreement adds a substantial number of additional obligations on matters where the pre-existing conventions are silent or were seen as being inadequate. The TRIPS Agreement is thus sometimes referred to as a Berne and Paris-plus agreement.

2. *Enforcement.* The second main set of provisions deals with domestic procedures and remedies for the enforcement of intellectual property rights. The Agreement lays down certain general principles applicable to all IPR enforcement procedures. In addition, it contains provisions on civil and administrative procedures and remedies, provisional measures, special requirements related to border measures and criminal procedures, which specify, in a certain amount of detail, the procedures and remedies that must be available so that right holders can effectively enforce their rights.

3. *Dispute settlement.* The TRIPS Agreement makes disputes between WTO Members about the respect of the TRIPS obligations subject to the WTO's dispute settlement procedures.

Source: World Trade Organization. http://www.wto.org/english/tratop_e/trips_e/intel2_e.htm#generalprovisions.

righted work (e.g., a translation or sequel to a novel or a new version of the original software program).

In the United States, a copyright automatically is established as soon as an original work is fixed in a tangible medium of expression. The use of a copyright notice on the copyrighted work is no longer required, but it is still useful to put the public on notice that the work cannot be freely reproduced or adapted for further use without receiving the copyright owner's permission. Also, although registration of a copyright with the federal copyright office is not required for the copyright to be legally effective, registration is required before one can sue others for copyright infringement or collect any statutory damages and reimbursement of enforcement costs.

In the United States, the remedies for copyright infringement include an injunction to stop the infringement and/or the actual damages suffered by the copyright owner due to the infringement, and any profits of the infringer attributable to the infringement.

Statutory damages (at the copyright owner's election to recover in lieu of actual damages and profits) are also available.

An individual author of works created after January 1, 1978 (the effective date of the Copyright Act of 1976) has copyright protection for his or her life plus seventy years. When works are made for hire (such as by an employee of a corporation that will own the copyright), the duration of copyright protection is the earlier of 95 years from first publication or 120 years from creation.

SOFTWARE COPYRIGHTS

Copyright is a relatively quick and inexpensive means of protecting a specific work from being copied, distributed, or used to create a new work that is largely derived from the original copyrighted work. For example, reproducing a copyrighted software program, which would occur automatically when downloaded from a Web site, is not permissible without the owner's consent. The same would be true with respect to making any changes to the owner's software application.

However, to make certain that users of software owned by someone else are legally restricted in a manner that maximizes the software owner's business strategy for the software, users are normally required to enter into a software license agreement that grants them the right to run the software in object code (machine-readable) form, which may be limited to use on a particular device, at a specified location, or throughout the end user's business on any device. The broader the scope of use granted to the end user, the higher the license fee charged will normally be to the end user. Click-through agreements on the Internet in which a user clicks his or her assent to terms of use posted on the software owner's Web site are legally enforceable in the United States, the United Kingdom, and other major industrialized countries.

Object Code

Object code is machine-readable code that is difficult to reverse engineer, but it can be illegally copied and used beyond the permissible bounds without being easily detected by the owner. Thus, off-the-shelf software packages that do not require any ongoing maintenance or modifications by the original owner for continued use are susceptible to pirated copying. The more dependent the user is for ongoing maintenance and enhancements that only the owner who controls the software's source code can provide (i.e., human readable version of the underlying program), the less susceptible the software is to being effectively misappropriated. That is why software owners are very reluctant to disclose their source code to unaffiliated users.

Apple Versus Microsoft

In *Apple Computer Inc. v. Microsoft Corporation*, 35 F.3d 1435 (9th Cir. 1994), Apple lost its copyright infringement claim, which had been based on the defendants' use of graphical user interface elements that were similar to those in Apple's computers. One of the reasons was a license agreement that Apple had signed with Microsoft that the court believed had granted Microsoft the right to use specific display elements. However, the court also examined the elements not covered by the license agreement and found them unprotected by copyright either because (1) they lacked originality, since Apple had admitted that it had taken certain ideas for icons from Xerox, (2) the idea of an icon in a desktop metaphor representing a document stored in a computer program can only be expressed in a limited number of ways and, therefore, the underlying idea and its expression are essentially merged, or (3) when features of a program are indispensable, or at least standard, in the treatment of a given idea, they are more akin to ideas than unique expressions.

LIMITATIONS OF COPYRIGHTS

Copyright protection can be fairly narrow in scope. This form of intellectual property protects the particular expression of an idea, but not the underlying idea itself. Thus, inventions, concepts, discovered facts, or an innovative idea as such, are not protected by copyright.

Moreover, in order for a work to be copyrightable, its author must demonstrate that it reflects a modicum of creativity and originality. For example, putting together an alphabetical list of names and telephone numbers has been determined by the U.S. Supreme Court as lacking sufficient creativity and originality to qualify for copyright protection.[1]

And no matter how much time and resources someone devotes to programming a software application, for example, there would be nothing to prevent someone else from independently creating an application with very similar features and functionality, as long as there is no evidence that he or she had prior access to and used the other's software application in whole or in substantial part.

Also, under current U.S. law, the data contained within a database is not copyrightable, although the compilation and look and feel of the database as a whole might be copyrightable if done in a sufficiently creative and original manner. Because the data collected and contained in databases—sometimes at great expense—will not receive copyright protection, the European Union decided to adopt a special directive designed to extend protection to the data itself that are collected and placed in databases (EU Database Directive 1996). The EU directive adds a sui generis right specific to databases, which it defines as "a collection of independent works, data or other materials arranged in a systematic or methodical way and individually accessible by electronic or other means." The directive prohibits unauthorized use or extraction of the facts collected within the database where the database owner can demonstrate "substantial investment in either the obtaining, verification or presentation of the contents."

In the United States, there is no such legislation, although bills to create a specific new intellectual property right in the data contained in a database have been introduced into Congress. Until such legislation is passed, however, database owners have to rely on state laws against unfair competition, which are not as effective as targeted statutory protection.

There are also limitations on the right of the copyright owner to exclude others with respect to the copyrighted work. One is the First Sale Doctrine, which allows the owner of a lawful copy—for example, a book purchased at a bookstore—to sell, lend, or otherwise dispose of that one purchased copy only. He or she cannot make several copies of that book and sell or otherwise distribute those copies, however. One exception to the First Sale Doctrine is the prohibition against renting out a purchased computer software program without the software owner's consent. Moreover, as of the time this book went to press, the U.S. Supreme Court is grappling with a case that could define the extent to which the First Sale Doctrine is applicable to goods that are not manufactured in the United States.

Another limitation on the right of the copyright owner to exclude others with respect to the copyrighted work is the Fair Use Doctrine, which allows free use of limited portions of a work for certain, primarily noncommercial purposes, such as news reporting, commentary, or educational materials.

With respect to software, the "open source movement" was founded upon the notion that software should be freely available and "open" to modification and development—a philosophy that is at odds with the proprietary notion underlying intellectual property protection. Thus, open source software is by definition not subject to the usual copyright restrictions. We discuss "open source" software in more detail later in this chapter.

INTERNATIONAL COPYRIGHT PROTECTION

While we focus this discussion on U.S. copyright law, any varying requirements outside of the United States for copyright notices, registration of the copyright with government authorities, and so forth must be checked before relying on copyright protection in those jurisdictions. The United States belongs to the Berne Convention for the Protection of Literary and Artistic Works (1886), which provides reciprocity of the validity of each member's copyright laws in the jurisdictions of the other members. As of the first quarter of 2012, 165 countries belonged to the Berne Convention.

INDUSTRIAL BRANCH OF INTELLECTUAL PROPERTY

The other major types of intellectual property mentioned at the very beginning of this section, based on the WIPO list of intellectual properties, such as inventions, industrial designs, trademarks, service marks, commercial names and designations, and unfair competition, constitute what WIPO refers to as the "industrial property branch" of intellectual property (WIPO Industrial Property 4). Although not specifically listed, trade secrets would also be considered an example of the industrial branch of intellectual property.

PATENTS

A patent is an exclusive right granted with respect to an invention, which may be thought of as an idea reduced to a practical application or use. A patent may involve a product or a process—the key is that the invention provides a new and innovative way of doing something or offers a new and innovative solution to a problem. In the United States, patents, like copyrights, are also recognized in the Constitution (Article 1, Section 8, Clause 8) and by federal statutory law.[2]

PREREQUISITES AND CONDITIONS FOR PATENTS

In order to qualify for a patent, the applicant and the invention must first meet certain prerequisites and conditions. The prerequisites include a showing that the invention is of practical use or industrial application. It must be novel, in the sense that it must advance the body of existing knowledge (referred to as prior art) in its technical field. And it must be nonobvious, in the sense that it demonstrates an innovative step that, in WIPO's words, "could not be deduced by a person with average knowledge of the technical field" (WIP Industrial Property 6).

The final prerequisite is that the invention must fall within the scope of patentable subject matter as defined by national law, which varies somewhat from country to country. Many countries exclude from patentability such subject matter as scientific theories, mathematical formulas, plant or animal varieties, or discoveries of natural substances.

Thus, the algorithms or mathematics themselves used in developing and coding software are not in and of themselves patentable. Software that manipulates numbers, abstract concepts, or signals representing numbers or concepts, without practical application to something tangible, is not patentable.

In addition to the foregoing prerequisites for patent eligibility, the major condition for the right to receive a patent is the future patent owner's willingness to adequately disclose the patented invention to the public, in order to enable its dissemination to others who may be able to further develop the technology. The clock will be ticking for the patent owner while the person or entity owning the patent tries to exploit the patent commercially during the period of legal monopoly over the patented invention. After that period expires, it will be an open field for competitors, seeded by the patent owner's own disclosures that were made part of the public record of the patent when issued. For this reason, for example, major pharmaceutical companies have to keep replenishing their patented drug pipeline as they lose patented protections for their earlier invented drugs and face intense competition from generic drug manufacturers.

PATENT HOLDER'S RIGHTS

In the United States, as is true in many countries, a patent is granted by the patent office of the country in which the owner of the invention seeks to protect it. A patent application must be filed in each jurisdiction where patent protection is sought, containing the invention's title, the applicable technical field, and a description of the invention written in a manner clear and detailed enough that someone with ordinary skill in the field can understand it and carry it out.

The patent owner (or patentee) may practice or use the invention; sell (assign) his or her patent; or license others to make, use, or sell the patented invention in exchange for royalty payments. The patentee is also free to decide not to license its patents at all or to cross-license with other firms so that it can receive valuable patent licenses in return from those firms.

Anyone who is not authorized to make, use, or sell any product or service that relies on the patented invention is guilty of infringing the owner's patent and is subject to being stopped by a court injunction from engaging in such prohibited activities, as well as to being assessed with damages owing to the patent owner. There is today a fairly wide scope for awarding damages. Damages may be based on a calculation of lost profits from the infringement if that is determinable, but can be no less than a "reasonable royalty"—a phrase that is not defined in the governing statute.

RELATIVE BENEFITS AND COSTS OF PATENTING

Benefits

The decision to patent involves a balancing of the costs of patenting and the risks of having the patent challenged and invalidated against the benefits that successful patents provide to their owners, which exceed the scope of copyright protection alone.

Software developers originally relied on copyrights as the preferred means to protect their intellectual property rights because it was much cheaper and quicker to obtain this form of protection. However, copyrights protect only the particular expression of an idea (i.e., the software code as written), not the innovative features, functionality, or architecture of the software. Developers and investors have recognized that patent protection is more effective and broader than copyright protection in ensuring that competitors are excluded from appropriating the rewards of the invention's success.

The payoff in securing patent protection is that it provides the owner of the patent the ability to exclude others from making or having made, selling, using, or otherwise exploiting the patented invention without the patent owner's consent. Think of patents as valuable assets that have the potential to enhance the overall value of a business, particularly where the patented innovations are critical to the business's success in maintaining a competitive edge in the market.

Incentive for R&D. Patents provide incentives to individuals or enterprises to devote considerable time, money, and other resources to useful research, with some assurance that they will be able to reap a material reward for their marketable inventions from licensing revenues. In addition to licensing revenues, firms reliant on high technology pursue patents in the fields of their core competencies to maximize the private return from their exclusive exploitation of patents they own themselves in their own business. For example, financial services firms engaged in highly sophisticated computer program–driven trading do not simply rely on software that they can license from others because that software is presumably available to their competitors as well, denying them any first-mover advantages. That is why investment banks such as Goldman Sachs have been aggressive in applying for their own patents to protect their proprietary trading strategies whenever possible, although their strategies are likely to be affected by the recent restrictions imposed on obtaining business process patents, discussed below.

Pharmaceutical patents, by contrast, are easier to defend than software business method patents because of the nature of R&D performed, yielding patents that are more detailed in their claims (Schacht 2000). Also, patents in the pharmaceutical sector are estimated to raise the costs incurred by non-patent holders wishing to use the idea or invent around the patent by 40 percent, as compared to the 7 percent to 15 percent range for electronics (Mansfield, Schwartz, and Wagner 1981).

Opportunity for Strategic Alliances and Joint Ventures. A company may also be able to use the leverage created by its patents to entice partners to enter into strategic alliances or joint ventures that may not have been as possible without the patent. Start-ups or small-technology companies use their patents to attract financing from venture capitalists or from larger, more financially established companies. Patent owner A may decide to cross-license one or more of its patents in order to obtain the right to use patents owned by patent owner B, enabling it to create new products or services that would not have been possible without fear of a patent infringement lawsuit. Finally, if a patent owner does face a lawsuit alleging that it infringed someone else's patent, the patent owner may be able to use its patents defensively in a counterclaim or as part of a settlement entailing cross-licenses with the plaintiff.

Create Legal Monopoly. In short, patents create a legal monopoly for a finite time in an invention—usually, in the United States, for 20 years from the date of filing the application for a utility patent, such as a software patent, or for a plant patent. This provides more protection than a copyright, which is limited to the particular way of expressing the idea behind the invention. Having a portfolio of patents has the effect of legally preventing competitors from obtaining patents on the same technology or using that technology in the competitor's business without obtaining a license.

Costs

Expensive and Time-Consuming. However, while patents can play a significant role in securing legal protection for one's research and technological innovations, the patenting process can be expensive and time-consuming. Moreover, success is often uncertain. According to the 2007 American Intellectual Property Law Association Economic Survey of 2006 costs, the average patent attorney fees for preparing the original complex application was $12,000 plus an additional patent prosecution cost for each relatively complex amendment and supporting paper of $3,000. To that sum must be added search and filing fees of approximately $4,000 per patent application (Ventre 2011). Note that according to the U.S. Patent and Trademark Office, applications that were filed but never issued constituted roughly *35 percent of total filings.* This means that a firm can end up investing significant resources in the patent process with a 35 percent chance of failure.

Since, as we discussed in chapter 1, it is fairly easy for someone to invent around a single patent, high-tech firms often file multiple patent applications to build a wall around their inventions. See Table 4.2 for the Microsoft example of number of patents granted and patent applications filed from 1998 to 2011.

Filing in Other Countries. Legal, search, and filing fees are incurred to obtain a patent in each country in which patent protection is sought. There is no such thing as an international patent or reciprocal treatment among various countries of each other's patents such as the Berne Convention provides for copyrights. However, the WIPO-administered Patent Cooperation Treaty does provide for international cooperation with regard to filing, searching, and examination of patent applications and the dissemination of applicable technical information.

Table 4.2

Microsoft's Patents by Year of Application and Year of Grant, 1998–2011

Year	Patents by Year of Grant	Patents by Year of Application
Pre-1998	395	1474
1998	334	641
1999	352	762
2000	344	1061
2001	396	928
2002	499	1116
2003	499	1724
2004	629	2850
2005	746	3216
2006	1463	1801
2007	1638	1189
2008	2026	582
2009	2901	169
2010	3086	95
2011	2309	9

Source: U.S. Patent and Trademark Office (2012).

Note: Data for applications that were filed but never issued (roughly 35 percent of total filings) are not included in this report.

Defending Against Infringement. Patent litigation to enforce one's own patent or to defend against an infringement suit by a patent owner can also be very expensive. Critics of the patent process claim that the substantial risks associated with patent infringement litigation and the enormous expense incurred in defending against such lawsuits have forced firms accused of infringement into settlements on disadvantageous terms. The legal expenses alone in defending against a typical patent infringement suit can be in the millions of dollars. The fear of being the target of such litigation with uncertain outcomes, including injunctions and huge damage awards, has deterred firms from undertaking innovative projects that have the real prospect of inviting infringement challenges from an opportunistic plaintiff. Such a plaintiff can prevail without having to prove that the alleged infringer copied its design or features or even knew that a patent covering the technology existed. See Exhibit 4.2 for an example of an expensive, bet-the-farm patent litigation that ended in a costly settlement.

PATENT LITIGATION EXPLOSION

Bessen and Meurer (2005, 28) conclude in their empirical study, "The Patent Litigation Explosion," that there is a cumulative impact of the increase of patent litigation across diverse industries, which could "represent a growing disincentive to R&D that is not likely offset by growth in the number or value of innovations." Software patents contribute to this growth as they cross over and affect a variety of applications in different industries such as telecommunications.

All in all, based on their findings, Bessen and Meurer dismiss firm patenting rates, R&D spending, firm value, or industry composition as the primary explanations for

Exhibit 4.2 **NTP, Inc. v. Research in Motion**

Perhaps one of the most notorious examples of a patent litigation, with nearly catastrophic effects on the defendant, involved the high-technology provider of the BlackBerry service Research in Motion ("RIM"). RIM's BlackBerry service, software, and devices allow individuals to send and receive information on a wireless basis, such as e-mail and company data, via handheld devices. RIM's own patent, which was granted in 2001, described a method of directing e-mail to wireless devices, while maintaining mailbox synchronization with a desktop e-mail system. The BlackBerry is a closed, single-vendor e-mail system, based on a set of proprietary protocols.

RIM was threatened with a potential shutdown of its service in the United States as a result of a patent infringement suit filed by NTP, Inc. ("NTP") against RIM in 2001. NTP is a Virginia-based patent holding company whose primary asset is a portfolio of at least 50 U.S. patents, which focused on inventions in the fields of wireless e-mail and RF Antenna design. NTP is an example of a company that regularly goes to court to aggressively enforce its patent ownership rights to the limit, even though it does not manufacture, sell, or use the patented invention itself in any product or service and does no technology research of its own.

NTP was highly successful in its lawsuit against RIM, winning a jury verdict and obtaining a ruling by a federal judge that granted an injunction which, when put into force, would prevent RIM from making, using, or offering to sell handheld devices, services or software in the United States that were deemed to infringe NTP's patents. RIM appealed all of the findings of the court and got a stay of execution pending the outcome of the appeals. However, after fits and starts at negotiations to settle the lawsuit, and with its appeals exhausted, RIM faced the prospect of an immediate shutdown of its service once the injunction was actually put into force (*NTP, Inc. v. Research in Motion* 2006).

RIM had some success in getting the U.S. Patent and Trademark Office to reverse its earlier findings regarding the validity of NTP's patents and to issue preliminary rejections of most of the applicable patents. RIM also sought to assure its major customers that it had developed a software work-around that would not infringe the NTP patents. However, all this was too little too late. After the federal judge in charge of the case ignored the patent office's preliminary decision to invalidate some of NTP's patents that had previously been approved, and issued an ultimatum that effectively put RIM's back to the wall, RIM finally reached a settlement agreement with NTP for $612.5 million.

the rapid rise in patent litigation. Instead they conclude that a more favorable climate for legal challenges against patent infringers has been the predominant driver of the rapid growth of litigation.

Supreme Court Rulings to Curb Runaway Litigation

In an effort to curb the threat of runaway patent litigation that can suppress real innovation, the Supreme Court has recently issued three important rulings.

In the *eBay v. MercExchange* (2006) case, the Court made it tougher for plaintiffs whose business is to license patents rather than use the patents themselves in their own products or services to wield the threat of injunction against those whom they accuse of infringing those patents.[3]

In the *KSR International Co. v. Teleflex, Inc.* (2007) case, a unanimous Supreme Court ruled in favor of applying more rigid standards for obtaining a new patent, excluding from eligibility for patent protection an advance that would be "obvious" to someone of ordinary skill in the field. The Court said that "granting patent protection to advances that would occur in the ordinary course without real innovation retards progress. . . . When there is a design need or market pressure to solve a problem and there are a finite number of identified, predictable solutions, a person of ordinary skill in the art has good reason to pursue the known options within his or her technical grasp. If this leads to the anticipated success, it is likely the product not of innovation but of ordinary skill and common sense."[4]

Finally, the Court ruled in 2010 that patents for business methods need to be more selectively granted than they have been over the last several decades. The case, which may particularly impact software programs that embody business ideas, involved a patent for a method of hedging in commodities trading. Quoting another of its opinions, the Supreme Court said that abstract ideas, laws of nature, and physical phenomena are "part of the storehouse of knowledge of all men . . . free to all men and reserved exclusively to none."[5]

BUSINESS METHOD PATENTS

State Street Bank & Trust Company v. Signature Financial Group, Inc., 149 F. 3rd 1368 (Fed. Cir. 1968), involving a patent for a mutual fund data-processing system, was a landmark case decided by the federal court specializing in patent issues (the U.S. Court of Appeals for the Federal Circuit) because it opened the floodgates to the filing and issuance of a category of patents known as business method patents. These are patents that disclose and claim new methods of doing business such as financial products, software used for processing financial transactions, Amazon.com's "1-Click" patent, Priceline's method, and the use of "buyer-driven commerce" on the Internet, and so forth.

The court applied a liberal standard for the granting of such patents—it was sufficient to show that the invention produces "a useful, concrete and tangible result" in order to be eligible for a patent.

As shown in Table 4.3, the U.S. Patent and Trademark Office has provided statistics for business method patents (Class 705), showing increases in the number of applications filed and patents issued.

MACHINE-OR-TRANSFORMATION TEST

But a 2008 opinion by the same U.S. Court of Appeals for the Federal Circuit (affirmed by a subsequent U.S. Supreme Court opinion) may well close the floodgates by toughening the standard for obtaining a business method patent. The appeals court decision, in the case known as *In re Bilski*, 545 F.3d 943 (Fed. Cir. 2008), which was affirmed by the Supreme Court in 2010, upheld a ruling made by the Board of Patent Appeals and Interferences that denied a patent for a method of hedging in commodities trading. This decision sharply cut back on what the appeals court's prior State Street decision had permitted. It did so by saying that it was no longer enough to show that the invention produces "a useful, concrete and tangible result" for patenting purposes. Instead, the court applied what is called the "machine-or-transformation test." Either the business method claim must be tied to a particular machine (i.e., specific computer device) or it must transform an article into a different state or thing.

As is often the case with intellectual property law, this test—particularly the "transformation" part—is easier to recite than to apply in real life because it is so esoteric in nature. However, the appeals court provided some guidance with examples. It said that simply manipulating abstract concepts such as legal obligations or business risks, including a process enabling the exchange of only options, would not qualify for patent protection. It is all too abstract in that it refers only to legal transactions and does not involve the transformation of any physical object or substance, or an electronic signal

Table 4.3

Class 705 Application Filings

Fiscal Year	Class 705 Serialized Filings	Class 705 CPA-RCE-R129 Filings	Class 705 Total Filings	Class 705 Issues
2002	6,774	626	7,400	494
2003	6,387	1,310	7,697	486
2004	6,681	1,731	8,412	291
2005	6,970	2,057	9,027	711

Source: U.S. Patent and Trademark Office (January 2009).

representative of any physical object or substance, into a different state or thing. A mathematical algorithm representing a mental process of performing certain mathematical calculations not tied to any specific computer device, identifying those transactions that the calculations have revealed would hedge each other's risks, and performing the post-solution step of consummating those transactions may well be a business process, but is not sufficient to qualify for a business method patent anymore.

On the other hand, for example, if X-ray attenuation data (which is representative of something physical and tangible, such as the structure of bones, organs, and other body tissues) is produced in a two-dimensional field by a computed tomography scanner, "the transformation of that raw data into a particular visual depiction of a physical object on a display" was enough to qualify for a patent. The court noted that as long as the data represented an underlying physical object, the electronic transformation of the data itself into a visual depiction was sufficient to qualify for a patent.

The Supreme Court, in affirming the appeals court decision, said that while the machine-or-transformation test was useful, it was not the sole test for patent eligibility as the appeals court had assumed. The Supreme Court concluded that it would have been sufficient to reject the patentability of the concept of hedging risk and the application of that concept to energy markets on the grounds that it was an abstract idea that did not meet the "new and useful" standard for patentability. As the Supreme Court pointed out, relying exclusively on the machine-or-transformation test "would create uncertainty as to the patentability of software, advanced diagnostic medicine techniques, and inventions based on linear programming, data compression, and the manipulation of digital signals" (*Bilski et al. v. Kappos*, 130 S.Ct. 3218, 3227[2010]). But at the same time, the Court recognized the need for more balance in the granting of patents.

Future cases will hopefully provide more guidance on where the balance will be struck. However, while the Supreme Court left the door open for granting some business method patents in the future, it is fairly clear that the heyday for business methods patents is over. It remains to be seen how innovators in this field will adapt.

TRADEMARKS

Trade names, marks, drawings, signs, logos, and the like are primarily used for branding and marketing purposes in order to distinguish the goods or services with which they are used. As shorthand, we will refer to all these particular forms of intellectual property collectively as trademarks. In the United States, trademarks are protected under federal statute.[6] Registered trademarks have a set term for protection (usually ten years) but this term is renewable.

Trademarks provide the owner the legal right to tag particular goods or services as those produced or provided by a specific person or firm. For example, "DELL" and the Apple logo are trademarks that identify goods (computers and computer-related products and services) as coming from Dell, Inc., and Apple Computer, Inc., respectively. Companies invest millions of dollars and other substantial resources in building up their brand recognition, which can enhance their good will and their competitive edge in the marketplace. They want to make sure that these advantages are not diluted by other products or services (including knockoffs that free-ride on the real trademark owner's investment) that carry the same or confusingly similar marks or names. Without this protection, the brand owner may suffer the loss of potential customers as well as reputational harm.

Trademarks give their owners a legal tool to protect their brand by giving them the exclusive rights to (1) use the trademark to identify their goods or services and prevent others from using and marketing the same or a confusingly similar trademark for the same or similar goods or services, and (2) grant others the right to use the trademark via license or franchise agreements as a source of revenue. In the United States, the remedy for a trademark infringement can be an injunction restraining the defendant from dealing in the infringing products and/or damages based on the plaintiff's lost profits, and injury to reputation, as a result of the infringement or the defendant's profits resulting from the defendant's sales of the infringing product.

COMPLEMENTARY MARKETING ASSETS

Like patents, trademarks can be a valuable business asset—in this case closely coupled with the firm's complementary marketing assets and strategies. See Exhibit 4.3 for a unique marketing strategy being used by the leading computer-based social network Facebook, which leverages trademarking to help its users.

In obtaining financing for high-tech firms or in setting the purchase price for a firm in an acquisition or merger, the value of the brand is often a critical component. Since without strong trademark protection a brand's value can be diluted by others' use of the same or confusingly similar branding, due diligence performed prior to the close of the purchase or merger of a business includes an examination of the level of trademark protection.

THE "WELL-KNOWN" TEST

If the trademark is considered sufficiently "well known" by the "relevant sector" of the public in a given country, it may even be protected against confusingly similar marks that are used on unrelated, dissimilar goods or services. In the United States, the Lanham (Trademark) Act grants to owners of "well-known" trademarks the right to seek injunctive relief against another person's use in commerce of a mark or trade name if such use causes dilution by blurring or tarnishing the distinctive quality of the famous trademark.

Exhibit 4.3 **Facebook Enables Its Users to Promote Their Trademarks**

Facebook, the leading social network which settled a trademark dispute over its own registered trademark, is allowing participants in its network to register personalized URLs (referred to as usernames) to direct interested parties to their Facebook pages (e.g., facebook.com/company name). What this does for businesses is to allow them to incorporate their registered trademarks into their usernames and reach the vast Facebook user audience, which is more than 600 million worldwide.

This service can generate even more "eyeballs" for Facebook, which enhances its own brand and value to advertisers. But what makes the service unique is that it offers businesses a ready-to-use complementary marketing asset with little additional investment required of their own.

Of course, like every good thing there are potential pitfalls. As an article from the *National Law Journal* posted on Law.com reports, "Lawyers note that while the new usernames will be a good marketing tool for companies to promote their products on Facebook and direct users to their Web sites, cybersquatters could snatch the trademark first and exploit it." Brian Fergemann, a partner and intellectual property attorney at Chicago's Winston & Strawn, was quoted in the article as saying that "It is a great tool. It's kind of expanding the ability of people to use Facebook as a platform for advertising ... but at the same time, it creates this potential opportunity for people to squat on other people's trademarks."

Market-savvy firms in today's Internet-driven society will do well to register their personalized usernames incorporating their trademarks for both offensive advertising purposes and for pre-emptive defensive purposes to stop someone else from using the name first, which would then entail potentially expensive litigation to stop.

Source: http://gold.insidenetwork.com/facebook/facebook-stats/ and Baldas (2009).

Well-known trademarks are also protected internationally under the Paris Convention for the Protection of Industrial Property (1883) as amended (Paris Convention), and Article 16 of the Agreement on Trade-Related Aspects of Intellectual Property Rights.

WIPO lists the following guidelines to be examined in determining whether a trademark meets the test of being sufficiently well known to qualify for this extra protection and defines the "relevant sector" of the public:[7]

1. The degree of knowledge or recognition of the mark in the relevant sector of the public
2. The duration, extent, and geographical area of any use of the mark
3. The duration, extent, and geographical area of any promotion of the mark
4. Including advertising or publicity and the presentation, at fairs or exhibitions, of the goods and/or services to which the mark applies
5. The duration and geographical area of any registrations, and/or any applications for registration, of the mark, to the extent that they reflect use or recognition of the mark
6. The record of successful enforcement of rights in the mark, in particular, the extent to which the mark was recognized as well known by competent authorities
7. The value associated with the mark

Relevant sectors of the public include:

(i) Actual and/or potential consumers of the type of goods and/or services to which the mark applies
(ii) Persons involved in channels of distribution of the type of goods and/or services to which the mark applies

(iii) Business circles dealing with the type of goods and/or services to which the mark applies

BENEFITS AND LIMITATIONS OF TRADEMARKS

In addition to providing brand protection for trademark owners, trademarks also benefit consumers by facilitating their choice in the marketplace when buying certain products or using certain services. The trademark helps the consumer to more readily identify a product or service, which he or she associates with the trademark from advertisements, and so on. Customers also have come to rely on trademarks associated with franchises for quality assurance, based on the quality standards that the franchiser imposes on its franchisees to maintain the reputation of the brand.

One limitation that trademark owners face is territorial scope. The owner of a registered trademark has an exclusive right to use the mark only within the territory within which the rights apply, unless the owner can meet the standards of a "well-known" trademark as noted earlier. The owner can prevent unauthorized third parties from using the mark, or a confusingly similar mark, in that territory so as to prevent consumers from being deceived into purchasing a product or service that they did not intend to purchase. A trademark must be registered in each country where protection is sought. Otherwise, it can be freely used and subject to some protection afforded to unregistered trademarks in some countries that is not as reliable as protection for registered trademarks. To facilitate multiple registrations of a trademark, WIPO administers a system of international registration of marks under the Madrid Agreement Concerning the International Registration of Marks and the Madrid Protocol. An owner of a registered trademark in one member country of these treaties can obtain an international registration that has legal effect in the other member countries. As of September 2010 there were 83 Contracting Parties to the Madrid Protocol.

DOMAIN NAMES

Domain names are Internet addresses used to find Web sites, often with names that identify the owner in much the same way as trademarks do. Think of Amazon.com, for example, which has come to be associated in online customers' minds with the large online store for the sale of books and other goods.

In fact, the U.S. Patent and Trademark Office will not register a domain name that would not pass muster under the criteria for approving a trademark. Thus, for branding and legal protection purposes, it is advisable for firms to use their domain names as they would a trademark in order to ensure retention of the name.

Cybersquatting

In order to deter the use of someone's trademark in another's domain name—known as cybersquatting—it is illegal in many countries for an unauthorized person to register the trademark of another firm or individual as that person's domain name.

Remedies may include requiring the cybersquatter to turn over the domain name to the trademark owner and pay damages.

Sometimes, there is a blurry line between using Internet addresses as a means to call attention to critical commentary about a company and misusing the company's trademark. For example, an investment adviser has registered the domain name goldmansachs666.com and has used it for a blog that criticizes Goldman Sachs's practices. Despite the disclaimer on the blogger's Web site that it is not affiliated with or approved by Goldman Sachs, the investment bank's counsel wrote the blogger a letter warning him that "Your use of the mark GOLDMAN SACHS violates several of Goldman Sachs's intellectual property rights, constitutes an act of trademark infringement, unfair competition and implies a relationship and misrepresents commercial activity and/or an affiliation between you and Goldman Sachs which does not exist and additionally creates confusion in the marketplace." In response, the blogger brought a lawsuit against Goldman Sachs to establish his right to use the domain name as part of a legitimate protest campaign against Goldman Sachs and that he is not in violation of Goldman Sachs's trademarks. Courts have in the past been sympathetic to noncommercial protest sites like this as opposed to sites that trade on a rival's name to create confusion in the marketplace and either divert business away from the rival or hurt its commercial reputation for anticompetitive purposes ("Goldman Sachs threatens blogger" 2009). See Exhibit 4.4 for another example.

Trade Secrets and Unfair Competition

A "trade secret" refers to the intellectual property right (separate and apart from patents or copyrights) protecting proprietary financial, business, scientific, technical, economic, or engineering information, including patterns, plans, compilations, program devices, formulas, designs, prototypes, methods, techniques, processes, procedures, programs, or codes, whether tangible or intangible, and whether or how stored, compiled, or memorialized physically, electronically, graphically, photographically, or in writing, if the following two conditions are met:

1. The owner of the proprietary information has taken reasonable measures to keep such information *secret,*
2. The information derives *independent economic value, including competitive advantage,* from not being generally known to, and not being readily ascertainable through proper means by, the public.

Trade secret protection is particularly important for high-tech firms seeking to protect their proprietary know-how, technology, and information that does not qualify for patenting but may constitute a critical input into creating a patentable invention.

Enforcement of Trade Secrets

Until 1996, trade secret owners in the United States had to look for protection primarily from state law, which provided remedies for "misappropriation" of a trade secret

Exhibit 4.4 **Trademark/Domain Name Case Example**

The following excerpt from a federal appeals court provides an idea of how courts will look at a person's use or abuse of someone else's trademark in a domain name:

SILVERMAN, Circuit Judge:

"Defendant Michael Kremer was dissatisfied with the hair restoration services provided to him by the Bosley Medical Institute, Inc. In a bald-faced effort to get even, Kremer started a website at www.BosleyMedical.com, which, to put it mildly, was uncomplimentary of the Bosley Medical Institute. The problem is that 'Bosley Medical' is the registered trademark of the Bosley Medical Institute, Inc., which brought suit against Kremer for trademark infringement and like claims. Kremer argues that noncommercial use of the mark is not actionable as infringement under the Lanham Act. Bosley responds that Kremer is splitting hairs. Like the district court, we agree with Kremer. We hold today that the noncommercial use of a trademark as the domain name of a website—the subject of which is consumer commentary about the products and services represented by the mark — does not constitute infringement under the Lanham Act."

But then the court ruled against the defendant Kremer under the Anti-cybersquatting Consumer Protection Act ("ACPA"), at least on a preliminary motion to dismiss the case altogether. It seems that Kremer allegedly may have been trying to extort a very high price from the plaintiff to buy the defendant's domain name:

"Cybersquatting occurs when a person other than the trademark holder registers the domain name of a well known trademark and then attempts to profit from this by either ransoming the domain name back to the trademark holder or by using the domain name to divert business from the trademark holder to the domain name holder... Allowing a cybersquatter to register the domain name with a bad faith intent to profit but get around the law by making noncommercial use of the mark would run counter to the purpose of the Act... the district court erred in granting summary judgment for Kremer on the ACPA claim. It remains to be seen whether Bosley can establish that Kremer registered the domain name in bad faith or can authenticate other letters that Bosley alleges were written and sent by Kremer."

Source: Bosley Medical Institute v. Kremer (2005).

(i.e., an acquisition, disclosure, or use of a trade secret by "improper means" such as theft, bribery, misrepresentation, breach, or inducement of a breach of a duty to maintain secrecy, or industrial espionage through electronic or other means. Now there is also the federal Economic Espionage Act of 1996, which imposes criminal and civil penalties for the theft or misappropriation of trade secrets in certain circumstances, by domestic and foreign entities.[8]

The Goldman Sachs Case

To provide an example of how trade secret enforcement can come into play, a former computer programmer at Goldman Sachs was alleged to have stolen Goldman Sachs's proprietary "black box" computer programs that the investment bank uses for its own and its clients' trading accounts. These programs constituted trade secrets. According to a report on this incident, "their value, experts say, could be incalculable" (Bowley 2009). The programmer was arrested by federal agents and accused of theft of trade secrets and transporting them abroad. The accused programmer allegedly used his desktop computer at his Goldman Sachs workplace to upload a stream of code to a Web site hosted by a Germany-based server and then downloaded the software files to his home computer, laptop computer, and a memory device.

In December 2010, the computer programmer was found guilty of theft of trade secrets and transportation of stolen property. Although his criminal conviction under the statutory language of the Economic Espionage Act was eventually overturned by the U.S. Court of Appeals for the Second Circuit, serious immediate economic injury to Goldman Sachs's proprietary trading may nevertheless have been prevented by the prompt legal action taken to enforce Goldman's trade secret rights.

Perhaps even more important than reliance on legal protections, Goldman's mitigation of its damages also depended heavily on its security systems, which detected the alleged theft in the first place, and on the significant complementary marketing assets that the alleged thief did not have access to but which were essential to optimizing the value of the software. These included its databases and electronic links to its customers.

UNFAIR COMPETITION LAWS

"Unfair competition" laws are closely linked with trade secret and trademark protections. They are not to be confused with antitrust laws, which deal with monopolization by a dominant firm and with collusion by competitors to the detriment of an efficient, competitive market. (Antitrust laws are discussed later in this chapter.) Unfair competition laws are aimed at protecting individuals and businesses from unfair or deceptive practices in industrial or commercial matters that include:

1. Misuse or unauthorized disclosure of someone else's secret or confidential information without the rightful owner's permission—related closely to trade secret intellectual property protection
2. Acts which may create confusion in the marketplace with the business, goods, or services of a competitor—related closely to misuse of someone's trademark and diluting the value of the brand
3. Making false and disparaging statements about a competitor, which can dilute the value of the brand by creating negative associations with the targeted firm's trademarks
3. False advertising

The Paris Convention for the Protection of Industrial Property requires its member countries to provide protection of industrial property against unfair competition. In the United States, unfair competition is primarily dealt with as a matter of state common (case) law; although some states have legislation that covers this subject area as well. At the federal level, the Lanham (Trademark) Act and the Economic Espionage Act of 1996 may also provide some relief.

Unfair competition law can be used by an aggrieved party who owns a trademark but has not yet registered it. But compensable injury based solely on unfair competition is difficult to prove. That is why it is strongly recommended that whenever national laws or international treaties allow for registration of intellectual property rights, registration should be pursued so that the owner of the trademark, for example, can invoke the statutory protections.

OPEN SOURCE SOFTWARE

BACKGROUND

The "open source movement" was founded upon the notion that software should be freely available and "open" to modification and development—hence the label open source software, or OSS for short. The Linux operating system is a prime example of OSS that was developed to counter Microsoft's monopoly in proprietary software operating systems. As the development of OSS evolved under the leadership of such organizations as the Open Source Initiative, open source software began to serve the needs of businesses while still ensuring the wide dissemination of software without restrictions on further innovative developments from multiple sources.

OSS programs are distributed subject to the terms and conditions of different forms of license agreements. The most common form of OSS licenses is the GNU General Public License (GPL), which is now in its third iteration.[9]

While these licenses give users certain freedom with respect to the use and modification of the subject programs, they also may contain certain affirmative obligations—sometimes referred to as "copyleft" terms. The most important terms include the following (Rao, Klein, and Chandra 2011):

- Any distribution to third parties or publication of a proprietary product that contains or is derived from open source software must be made available at no charge, assuming that the resulting product is distributed or published as one integral "whole" work. In other words, if a developer combines OSS code with the developer's own proprietary code or derives a proprietary software product from the open source code, that developer will not be able to impose a licensing fee for the resulting combined product (referred to here as a "hybrid product"). The whole work must be made freely available on the terms of the applicable license. What is considered a "whole work" will likely turn on a number of technical issues such as how closely the programs interact, how they are linked, and how the proprietary program loads with the underlying open source kernel (which is the underlying portion of the operating system that interacts the most closely with the hardware on which the operating system runs).
- The distribution of the hybrid product in object code form must be accompanied by the complete corresponding source code for the product or an opportunity to access such source code.
- All patents considered essential to implementation and use of the hybrid product must be either licensed under royalty-free terms for unrestricted use, or be covered by a promise of nonassertion when utilized in connection with the open source software. If the developer initiates a patent litigation against any third party alleging that the code distributed under the open source license agreement infringes the developer's patents, all licenses to any particular code granted to the developer under the open source license terminate—including with respect to the developer's rights in the open source code embodied in its hybrid product. This is a backdoor way of forcing the developer to choose between forgoing an enforcement action to stop or punish use of its own patents in open source software (and in effect give up the ability to profit directly from the patent monopoly itself), or sue for patent infringement and stop its own beneficial use of that same open source software.

THE CO-EXISTENCE AND CONVERGENCE OF OPEN SOURCE AND PATENT REGIMES

At first glance, the open source ethos of promoting free sharing of innovations would appear to conflict sharply with the rationale of patenting, which is to protect proprietary innovations from misappropriation and to internalize the return from R&D as much as possible.

If this were so, one would expect to see very divergent paths followed by the open source development community on the one hand, whose members eschew patent protection for their work, versus major software firms on the other hand, who wish to maintain the proprietary character of their products and rely for that purpose on patent and other traditional intellectual property protections as much as possible. However, to a significant degree, the opposite has happened. Even major high-tech incumbent firms have seen the benefit of sharing in unrestricted use of a common pool of copyrighted and patented works for which they are willing to waive their patent enforcement rights against other contributors to the common pool. These firms include IBM, Sony, Philips, Red Hat, and Google, which belong to an organization called the Open Invention Network. IBM and Red Hat are major vendors of products that incorporate open source software. They make a significant portion of their money from the service side of their business.

In his article concerning the commercialization of open source software, Mann (2006) discusses how established firms, which want to benefit from access to open source software in order to use the code as interoperable commoditized building blocks for their own products and services, have become big players in the open source arena. As a result, they have contributed some of their own patents to the common pool. All contributors to the pool—large and small—agree to cross-license open source–related patents to all other contributors at no charge. This cumulative shared portfolio then gives the participating patent owners the ability to negotiate access to other patented technologies and defuse potential litigation through further cross-licensing arrangements. The strategy, to quote Mann (2006), is to use "combined patent portfolios to create fences around some open source technologies." They provide some measure of protection for all contributors from the uncertainty that an investment of resources in an innovative project could yield nothing more than a ripe target for a patent infringement suit. Instead, they can freely collaborate on more innovative projects.

This does not mean that open source software participants are immune from patent infringement litigation. A lawsuit was filed in October 2007 against Red Hat and Novell by a nonparticipating patent-holding firm, for example, which claimed that its patents were infringed by various Linux-based products. And Microsoft, which did not historically participate in the common pool under the GPL open source license, has threatened to bring patent infringement suits against Linux vendors such as Red Hat. Parloff (2007) notes that Microsoft has claimed that approximately 235 of its patents are allegedly being infringed by open source code. On the other hand, Microsoft found itself in potential violation of GPL license terms when it incorporated a third party tool containing open source code as part of its Windows 7 product, which it took steps to rectify by bringing the tool under the terms of a GPL license.

Microsoft has pursued some level of coexistence with portions of the open source community. For example, it has chosen to team up with Novell (which, like Red Hat, is a participant under the GPL open source license) to share their respective patent portfolios on a bilateral basis. Moreover, Microsoft's Shared Source Initiative, which gives licensees the right to review and in some cases modify its source code for several of its platform products, represents an example of a trend on the part of firms producing proprietary software seeking to open up their own platforms and obtain in return the benefits of the open source model (Hahn 2002; Rao, Klein, and Borg 2006).

THE SPREAD OF INNOVATION BY LEAD USERS FROM OPEN SOURCE TO OPEN WIRELESS PLATFORMS, BUT CONVERGENCE WITH PATENT REGIME LAGS BEHIND

Google is a major user of open source software in its internal systems as well as in its online services, which generate substantial revenue for Google from Internet advertising. As such, Google has a strong incentive to contribute to OSS innovation. Google fits von Hippel's (2005) definition of lead users as members of a user population having two distinguishing characteristics: (1) they are ahead on an important market trend and are currently in need of solutions long before many users need them; and (2) they expect to derive relatively high benefits from obtaining a solution to their needs, and so they innovate. Morrison, Roberts, and Midgley (2004) found significant correlation between the two defining characteristics of lead users and the likelihood that they will innovate.

Google's encouragement of open source solutions is evidenced by innovations it has helped promote in the wireless telecommunications market. Google has spearheaded a group of more than thirty large corporations—including high-tech companies and mobile carriers—to join an organization called Open Handset Alliance. The chip makers Intel and Qualcomm, handset maker Motorola, wireless carriers T-Mobile and Sprint Nextel, and e-commerce provider eBay are among the members who committed to fostering an open Linux-based platform that allows the development of diverse but interoperable mobile handset applications. The resulting platform is called Android. The objective is to accelerate the ability of wireless customers to tap into the Internet for software, content, and services the way they can from their personal computers.

Just like their OSS counterparts with whom Google has also associated, participants in the Open Handset Alliance are contributing intellectual property to a common pool on a royalty-free basis. Would the result be more innovative wireless offerings on handset devices as firms competing at the top application layer know that their features will be interoperable with other features built on the same open platform, and will they be able to innovate under the protective umbrella of the contributed patents?

Android smartphones are already available in the marketplace. Consumers have benefited from the additional choices available to them. Makers of handsets utilizing the Android operating system have also benefited. They "represented the majority of the fastest-growing firms among the Top 10 smart phone brands in the second quarter [of 2010]," according to a statement released by the mobile and wireless research firm iSuppli Corp (Teng 2010). The research firm International Data Corporation reported that, during the third quarter of 2012, 75 percent of the smartphones shipped worldwide contained the Android operating system (2012).

However, as the market for innovative smartphones and applications has grown exponentially, some leading firms are relying on their own proprietary patented technologies and relationships with third party application developers to vie for market share, rather than rely on an open source platform. Apple is the most prominent example of a high-tech firm that has followed this strategy, relying on its proprietary operating system platform with user-friendly interfaces and a rich eco-system of compatible third party applications that customers desire.

Forrester analyst Charles Golvin (2012) describes the battle for customer loyalty this way in commenting positively on Apple's introduction of its iPhone 5 in 2012: "Apple still outpaces the competition when it comes to the entire package—the new iPhone unites significant improvements in industrial design, imaging, audio, and connectivity, along with the wealth of new capabilities that iOS6 enables. . . . But make no mistake, this is not about the iPhone 5 versus the Samsung Galaxy S III or the iPad versus the Kindle Fire HD; this is about customers'attachment to the larger ecosystems that those devices inhabit. Amazon, Apple, Google, and Microsoft all aim to translate customers' investments—of money, information, personalization, and social connections—into a gravitational field of loyalty so powerful that few customers will ever attain escape velocity."

Apple is not only fighting its battles with its competitors in the smartphone marketplace. It is also aggressively enforcing its patent rights in court, including against open source Android-based firms, as illustrated in Box 4.1.

The convergence of open source and patent regimes has not yet taken hold in the smartphone market to the extent it has in the software market.

PUBLIC POLICY IMPACTS ON USE OF INTELLECTUAL PROPERTY

This section reviews two significant public policy constraints on the use of intellectual property and proprietary information as part of a firm's marketing strategies. This involves antitrust and privacy laws.

ANTITRUST

Some Basics

Antitrust laws in the United States, the European Union, and other jurisdictions are aimed principally against (1) collusion between and among competitors (i.e., price-fixing, allocation of customers or territories, and group boycotts or exclusion of disfavored competitors), and (2) anticompetitive conduct by firms that dominate a market, such as predatory pricing or forcing a customer to buy a product he or she

Box 4.1

Apple v. Samsung

Apple and Samsung are currently the leading producers of smartphones. They have been embroiled in patent litigation, with each firm accusing the other of patent infringements.

On August 24, 2012, a federal jury in California reached a verdict in one such case, awarding Apple $1.05 billion in damages for Samsung's infringement of six Apple patents involving design and functional features such as the particular shape of the iPhone, the placement of onscreen icons and certain pinch-to-zoom and scrolling features.

What makes this case so potentially significant, aside from its direct impact on Samsung's bottom line if the verdict is not overturned on appeal, is the potential impact on the overall market for Android-based smartphones. That is because Samsung's devices are among the smartphones that utilize a variation of Google's Android operating system.

Apple has viewed with growing alarm the growth in market share of Android-based smartphones, which is larger in the United States and globally than the market share of Apple smartphones.

Steve Jobs, the late co-founder and chief executive of Apple, labeled Google's Android mobile operating system "grand theft" and vowed, "I'm willing to go thermonuclear war on this."

Apple has won round one. If the verdict is upheld on appeal, it will give Apple enormous leverage to slow the growth of devices using Google's competitive Android operating system or, alternatively, to demand major concessions in cross-licenses of patents from Google and smartphone manufacturing companies as the price for not pressing further infringement claims.

In addition, Microsoft may stand to benefit. It has developed its own version of an operating system for smartphones that is said to bear little resemblance to Apple's software, while entering into licensing agreements with some smartphone manufacturers whose use of the Android system without such licenses could infringe Microsoft's patents. The *Apple v. Samsung* decision may put more pressure on other manufacturers and on Google itself to enter into such licensing agreements with Microsoft in order to avoid having to fight a two-front war against both Apple and Microsoft.

Whatever its ultimate outcome, the *Apple v. Samsung* case illustrates the tension that still exists between firms like Apple, committed to a proprietary model of innovation, and firms like Google, committed to a more open source model for producing innovations. Smartphones represent the fastest growing market in computing technology and, with many billions of dollars at stake, a lodestone for patent litigation. Firms will find it increasingly necessary to accumulate portfolios of patents to use defensively as trades in cross-licenses as well as offensively to ward off would-be infringers. Meanwhile, consumers will be caught in the middle, with potentially fewer choices available in the marketplace as the specter of patent litigation chills innovation and scares off new entrants.

does not want as a condition for being able to purchase a desired product in which the firm has significant market power (known as tying).

In the United States, the Sherman Antitrust Act (Section 1) deals with agreements between and among firms in restraint of trade. The Sherman Antitrust Act (Section 2) deals with monopolization by a single dominant firm.

In the European Union, agreements between two or more firms that restrict competition are prohibited by Article 81 of the European Union (EU) Treaty. Firms in a dominant position may not abuse that position under Article 82 of the treaty.

Among the restraints that have been held per se unlawful are naked price-fixing, output restraints, and market division among horizontal competitors, as well as certain group boycotts. These acts are considered so inherently wrongful and anticompetitive that no defenses or justifications for them are permitted in court. There could be criminal as well as civil liability for per se violations of the antitrust laws.

In other cases, there is a balancing of anticompetitive effects against pro-competitive benefits in the affected markets under what is known in the United States as the "rule of reason" analysis. Other jurisdictions such as the European Union also engage in such a balancing of factors without explicitly using the rule of reason label for what they are doing.

As with other forms of private property, certain types of conduct with respect to intellectual property may have anticompetitive effects that the antitrust laws are intended to address. Intellectual property, including the right to exclude others from using the owner's intellectual property, is "neither particularly free from scrutiny under the antitrust laws, nor automatically suspect under them" (DOJ-FTC Antitrust Guidelines, Section 2.1 [1995]).

Relevant Markets for High-Tech Firms

With respect to high-tech firms in particular, there are three relevant markets that are defined for antitrust analysis purposes, as outlined in the DOJ-FTC Antitrust Guidelines. The first is not unique to high-tech firms—the market for the goods or services being sold by the firms involved. Intellectual property is often licensed, sold, or transferred as an integral part of a marketed good. An example is a patented product marketed with an implied license permitting its use, such as a software operating system embedded in computer hardware that makes the hardware run. In such circumstances, there is no need for a separate analysis of technology markets to capture relevant competitive effects.

The second market, however, more directly impacts high-tech firms than mass consumer product firms—this is the technology market, which consists of the intellectual property that is licensed separately from any underlying goods and services and "the technologies that are close enough substitutes significantly to constrain the exercise of market power with respect to the intellectual property that is licensed" (DOJ-FTC Antitrust Guidelines, Section 3.2.2 [1995]).

Related to the technology market is a third market that also has an impact on high-tech firms, known as the innovation market. An innovation market consists of "the research and development directed to particular new or improved goods or processes, and the close substitutes for that research and development" such as other research and development projects, technologies, and innovative products that are already being sold and may compete with innovations still in the R&D phase (DOJ-FTC Antitrust Guidelines, Section 3.2.3 [1995]). A separate delineation and analysis of innovation markets comes into play when "the capabilities to engage in the relevant research and development can be associated with specialized assets or characteristics of specific firms," such as a high-technology focus (DOJ-FTC Antitrust Guidelines, Section 3.2.3 [1995]).

Market Power

Market power is the ability to profitably maintain prices above, or output below, competitive levels for a significant period of time. Although the intellectual property right confers the power to exclude with respect to the specific product, process, or work in question, there are often sufficient actual or potential close substitutes for a given product, process, or work to prevent the exercise of market power. If a patent or other form of intellectual property confers market power, that market power does not by itself offend the antitrust laws if it is deemed the consequence of a superior product or business acumen. Nor does such market power automatically impose on the intellectual property owner an obligation to license the use of that property to others. However, where ownership of bottleneck facilities, such as the local network facilities of the incumbent local telephone exchange (including the intellectual property embodied therein), enables the strategic denial of access to a dominant firm's competitors, the result may be to deny "consumers the full benefits of technological progress that a dynamically competitive market would offer" (Shapiro 1999).

Licensing Arrangements Under the Antitrust Laws

To determine whether a particular restraint in a licensing arrangement is given per se or rule-of-reason treatment (or their equivalent in jurisdictions outside of the United States), the antitrust enforcers and courts assess whether the restraint in question can be expected to contribute to an efficiency-enhancing benefit for consumers. In general, intellectual property licensing arrangements promote such efficiency-enhancing benefits because they facilitate the combination of the licensor's intellectual property with complementary technological, production, and marketing assets owned by the licensee, which can result in lower cost innovative products that otherwise might not have been available as quickly in the marketplace. In this connection, certain license restrictions on what licensees can do with the licensor's intellectual property will not present an antitrust concern. As explained in the DOJ-FTC Antitrust Guidelines, Section 2.3 (1995):

> Field-of-use, territorial, and other limitations on intellectual property licenses may serve procompetitive ends by allowing the licensor to exploit its property as efficiently and effectively as possible. These various forms of exclusivity can be used to give a licensee an incentive to invest in the commercialization and distribution of products embodying the licensed intellectual property and to develop additional applications for the licensed property. The restrictions may do so, for example, by protecting the licensee against free-riding on the licensee's investments by other licensees or by the licensor. They may also increase the licensor's incentive to license, for example, by protecting the licensor from competition in the licensor's own technology in a market niche that it prefers to keep to itself.

However, not all intellectual property license restrictions will pass antitrust muster. One example of a licensing arrangement that would raise very serious anticompetitive concerns involves a patent licensing arrangement with multiple competing firms that imposes severe restrictions on the ability of the firms to compete with each other in markets for products not covered by the patent.

Another example of a licensing arrangement that might have anticompetitive effects involves what are called tying arrangements. Conditioning the ability of a licensee to license one or more items of intellectual property (the tying product) on the licensee's purchase of another item of intellectual property or a good or a service that the licensee may not want (the tied product) has been held in some cases to constitute illegal tying.[10] Although tying arrangements may result in anticompetitive effects, such arrangements can also result in significant efficiencies and pro-competitive benefits. Thus, antitrust enforcers and the courts would be likely to challenge a tying arrangement if: "(1) the seller has market power in the tying product, (2) the arrangement has an adverse effect on competition in the relevant market for the tied product, and (3) efficiency justifications for the arrangement do not outweigh the anticompetitive effects" (DOJ-FTC Antitrust Guidelines, Section 5.3 [1995]).

On the other hand, the fact that a patent, for example, confers a monopoly on the specific idea or invention covered by the patent does not automatically mean that the patent owner would have the power to monopolize a broader market containing reasonably close substitutes for the patented product or technology.

Package licensing or bundling—the licensing of multiple items of intellectual property in a single license or in a group of related licenses—can enhance efficiency when they combine what customers need in a usable product that depends on more than a single item of intellectual property. However, it could turn into a form of tying arrangement if the licensing of one product the customer wants to use on a stand-alone basis is conditioned upon the acceptance of a license of another, separate product that the consumer does not want.

Exclusive dealing arrangements may also have anticompetitive effects, particularly where they stifle innovation. For example, in the mobile device market, should a wireless carrier with significant market power in the national U.S. market be able to lock up certain mobile devices like Apple's iPhone under an exclusive deal? Does single-carrier control over the distribution of these devices have the potential of limiting the range of applications that will be run on the phones? But without such an exclusive arrangement, would the carrier be willing to order a sufficient number of handsets to incentivize the manufacturers to create innovative platforms for applications in the first place? These questions do not have settled answers yet from the courts.

CROSS-LICENSING AND PATENT POOLING

Cross-licensing involves the exchange of intellectual property rights between two or more persons. Patent pools are the sum total of intellectual property rights that are the subject of the cross-licensing.

Cross-licenses and patent pools are recognized by antitrust authorities as having pro-competitive benefits in that they help sellers provide consumers with better products and services at lower prices resulting from more efficient production and combining complementary inventions. They may help lower transaction costs and provide a means to counter the sort of burdensome patent infringement litigation that we discussed earlier in this chapter. At the same time, cross-licenses can have anticompetitive effects, too, including increased prices, cutbacks in production, and reduced innovation. This is particularly true when the cross-license is between firms that are competitors, whether in producing goods or services, licensing intellectual property, or in R&D. In the context of settling infringement litigation, settlements can reduce competition in markets with a small number of competitors to begin with or can involve an intellectual property asset that is fundamental to the ability of firms not part of the cross-licensing arrangement to compete in the affected markets. If, for example, the terms of cross-licenses as part of a settlement exclude the competitors of the parties to the cross-licenses from having access to a patented technology that is essential to their ability to conform to the technical requirements for a standardized interface, the anticompetitive effect of that exclusion will probably render it an antitrust violation.

Antitrust enforcement agencies and the courts analyze whether these cross-licensing or pooling arrangements hurt competition that either already existed or likely would have come into being without it. Using such arrangements as a device to fix prices or to divide territories or product markets between competitors is considered a per se violation of the antitrust laws without any further examination of the details of the markets involved or the rationales for the arrangements. In other cases, factors that will be weighed by antitrust enforcers and the courts in a rule of reason analysis (or its equivalent in non-U.S. jurisdictions) include the relationship of the intellectual property rights being combined. Are the patents complementary or do they actually cover competing technologies in a market where few if any alternatives exist? In the latter context, cross-licensing may be anticompetitive. Are members free to license their future patents independently of the pool, or is the pool the only way to get technology to the market? If the pool is the only alternative, what is the justification for this arrangement?

PRIVACY AND DATA PROTECTION

The online marketplace is growing at an "exponential rate," enabled by sophisticated technology allowing "online companies to collect, store, transfer, and analyze vast amounts of data from and about the consumers who visit their Web sites" (FTC 2000). The high technology of the Internet allows even the smallest companies that can afford to set up a Web site to gain a national and even an international marketing footprint if they so choose and to devise sophisticated target marketing strategies based on information about consumers, their interests, preferences, and communications patterns that they can collect electronically.

However, the very technological innovations that have sparked the huge increase in the collection and use of data, along with the myriad subsequent uses of this information that interactive technology makes possible, has also sparked increased concern about online consumer privacy. Who owns the information that consumers willingly or unknowingly provide to the Web site owners as they seek access to the site for informational purposes or to conduct transactions? What intellectual property rights to his or her own personal information does the consumer give up in the process?

Without deciding the ultimate question of what intellectual property rights the consumer retains or has given up with regard to his or her personal information in order to use a Web site collecting the data, legislatures and regulators in the United States and around the world have, to varying degrees, sought to clamp down on how such personal information can be used and how it must be secured by the Web site owner. Obligations are imposed on the enterprise in control of the data to maintain adequate technical, physical, and operations protections to ensure the security of the personal information. It is also responsible to ensure that any service provider processing data on behalf of the enterprise fully complies with the enterprise's instructions and all applicable laws and regulations. Also, to varying degrees, the enterprise in control of the information must notify the affected individuals about its data privacy policies and practices, obtain consent from the individual for certain kinds of uses and disclosures to third parties, and provide the individual with access and the ability to correct his or her personal information stored in the enterprise's electronic systems.

The Center for Democracy and Technology (CDT) has a Web site that discusses the various federal and state statutes in the United States, which are intended to protect individuals' privacy against both governmental and commercial intrusion and the security of the information.[11] Despite the Federal Trade Commission's recommendation to Congress to pass legislation containing comprehensive cross-industry privacy protections for consumers, Congress has not yet come up with such an all-encompassing law. So we are left with statutes specific to different industries or different types of information.

As categorized and listed by the CDT, the following lists federal statutes that define how governmental and commercial entities can collect, disclose, and use information in distinct areas:

Privacy of Communications

- The Electronic Communications Privacy Act (1986)
- Telecommunications Act (1996) Customer Proprietary Network Information

Children's Privacy

- Children's Online Privacy Protection Act (COPPA) of 1998

Privacy of Financial Information

- Fair Credit Reporting Act (1970)
- Right to Financial Privacy Act (1978)
- Taxpayer Browsing Protection Act (1997)
- Gramm-Leach-Bliley Act (1999)
- Fair and Accurate Credit Transactions Act (2003)

Privacy of Government Collections

- Computer Security Act of 1987
- E-Government Act of 2002
- Privacy Protection Act of 1980
- The Electronic Communications Privacy Act (1986)

Privacy of Medical Records

- Health Insurance Portability and Accountability Act of 1996 (HIPAA)

There are stricter privacy regulatory regimes outside of the United States that apply to all industry sectors and to personal data that a business collects about its employees, customers, and end users. The European Union (EU) Data Protection Directive is the best known of these. Taking effect in 1998, the purpose of the EU Directive was to mandate a core set of privacy-related requirements within all EU member states. EU countries also have extra protections for what they regard as highly sensitive personal information, such as medical records and data about an individual's racial or ethnic origin. Encryption of such data may be required.

The bottom line is that, just because highly sophisticated technology exists to help businesses more effectively market their products and services to consumers online, it does not mean that they are completely free, for any purpose whatsoever, to use personal data about their customers that technology enables them to collect. We can expect even more laws and regulations in the privacy and data protection arena as the legal system tries to keep up with technological change.

SUMMARY

In this chapter, we have surveyed the main types of legal protection for high-tech innovations under intellectual property laws, including copyrights, patents, trademarks, and trade secrets.

We observed that laws to protect intellectual property serve two primary purposes. The first is to grant legally recognizable rights to creators and other producers of the knowledge and information embodied in goods and services to control their use and distribution. The second purpose is to promote, as a matter of public policy, innovation and the dissemination and application of its results in order to stimulate technological change and the enhanced productivity that it produces.

However, we also noted that intellectual property laws alone are insufficient to ensure that the innovator reap as much of the economic value from the innovation as possible. Copyrights, for example, protect the particular expression of an idea, but not the underlying idea itself. Thus, inventions, concepts, discovered facts, or innovative ideas as such are not protected by copyright. A patent, which is an exclusive right granted with respect to an invention that protects an idea reduced to a practical application or use, provides broader protection for eligible innovations but is expensive to obtain as well as to uphold and enforce, particularly if the patent becomes the subject of litigation.

We discussed the "open source movement," which was founded upon the notion that software should be freely available and "open" to modification and development. While at first glance the open source ethos of promoting free sharing of innovations would appear to conflict sharply with the rationale of patenting, which is to protect proprietary innovations from misappropriation and to internalize the return from R&D as much as possible, we noted an increasing counter-intuitive trend among major high-tech incumbent firms in the software market particularly that seek the benefits of sharing in unrestricted use of a common pool of copyrighted and patented works for which they are willing to waive their patent enforcement rights against other contributors to the common pool. Established firms involved in software development and distribution, wanting to benefit from access to open source software in order to use the code as interoperable commoditized building blocks for their own products and services, have become big players in the open source arena. As a result, they have contributed some of their own patents to the common pool. All contributors to the pool—large and small—agree to cross-license open source–related patents to all other contributors at no charge. This cumulative shared portfolio then gives the participating patent owners the ability to negotiate access to other patented technologies and defuse potential litigation through further cross-licensing arrangements. We also observed that convergence of open source and patent regimes has not yet taken hold in the smartphone market to the extent it has in the software market.

The type of legal intellectual property protection that is most closely coupled with a firm's complementary marketing assets and strategies is trademark protection. Trademarks, as we observed, give their owners the legal tools to protect their brand by giving them the exclusive rights to (1) use the trademark to identify their goods or services and prevent others from using and marketing the same or a confusingly similar trademark for the same or similar goods or services, and (2) grant others the right to use the trademark via license or franchise agreements as a source of revenue.

Finally, we examined in this chapter two significant public policy constraints on the use of intellectual property and proprietary information as part of a firm's marketing strategies—antitrust and privacy laws.

KEY TERMS

antitrust
business method patents
copyrights
cybersquatting
intellectual property
lead users
machine or transformation test
object code
open source software
patent pools
patents
privacy
trademarks
trade secrets
"well-known" test

DISCUSSION QUESTIONS

1. What are the primary purposes of legal protection for intellectual property?
2. What does a copyright protect and what doesn't it protect?

3. What are the pros and cons of patenting software?
4. How do trademarks help a firm in the branding of its products?
5. What is "open source" software and how does it differ from patented software?
6. Why do some major high-tech firms with patented software support open source software as well?
7. To what extent has the open source movement affected the market for smartphones?

NOTES

1. *Feist Publications, Inc. v. Rural Telephone Serv. Co.*, 499 U.S. 340 (1991).
2. 35 U.S. Code Title 35—Patents Available at http://uscode.house.gov/download/pls/35T.txt.
3. *eBay Inc v. MercExchange, L.L.C.*, 547 U.S. 388 (2006).
4. *KSR Int'l Co. v. Teleflex, Inc.*, 550 U.S. 398 (2007).
5. *Bilski et al. v. Kappos*, 130 S.Ct. 3218, 3225 (2010) (quoting *from Funk Brothers Seed Co. v. Kalo Inoculant Co.*, 333 U.S. 127, 130, 68 S.Ct. 440, 92 L.Ed. 588 (1948).
6. Lanham (Trademark) Act (Title 15 of the U.S. Code).
7. Joint Recommendation Concerning Provisions on the Protection of Well-Known Marks adopted by the Assembly of the Paris Union for the Protection of Industrial Property and the General Assembly of the World Intellectual Property Organization (WIPO) at the Thirty-Fourth Series of Meetings of the Assemblies of the Member States of WIPO, September 20 to 29, 1999 at 6–7.
8. Title 18 of the U.S. Code.
9. According to the Web site devoted to what is called the GNU Project, "the name 'GNU' was chosen because it met a few requirements; first, it was a recursive acronym for 'GNU's Not Unix,' second, because it was a real word, and third, it was fun to say (or Sing)" http://www.gnu.org/gnu/gnu-history.html.
10. See, e.g., *United States v. Paramount Pictures, Inc.*, 334 U.S. 131, 156–58 (1948) (copyrights); *International Salt Co. v. United States*, 332 U.S. 392 (1947) (patent and related product).
11. http://www.cdt.org/.

REFERENCES

Baldas, Tresa. 2009. Facebook's imminent "username" registration raises risks of cybersquatting. *National Law Journal* (June). Available at http://www.law.com/jsp/article.jsp?id=1202431412892.

Bessen, James, and Michael J. Meurer. 2005. The patent litigation explosion. BUSL Working Paper 05–18. Available at The Social Science Research Network Electronic Paper Collection: http://ssrn.com/abstract=831685.

Bosley Medical Institute v. Kremer. 2005. 403 F.3d 672 (9th Circuit).

Bowley, Graham. 2009. Ex worker said to steal Goldman code. *New York Times*, July 7.

Directive No. 96/9/EC of the European Parliament and of the Council, of 11 March 1996 on the legal protection of databases.

Federal Trade Commission. 2000. "Privacy Online: Fair Information Practices in the Electronic Marketplace." Available at http://www.ftc.gov/reports/privacy2000/privacy2000.pdf.

Golvin, Charles. 2012. iPhone 5 strengthens Apple's leadership in the larger competition among ecosystems. *Forrester Blogs*. Available at http://blogs.forrester.com/charles_golvin/12-09-12-iphone_5_strengthens_apples_leadership_in_the_larger_competition_among_ecosystems.

Goldman Sachs threatens blogger over demonic name. 2009. *The Register* (April). Available at http://www.theregister.co.uk/2009/04/14/goldman_sachs_666_in_court/.

Hahn, Robert. 2002. Government policy towards open source software. *AEI-Brookings Joint Center for Regulatory Studies*. Washington, DC.

International Data Corporation Press Release. 2012. *Android Marks Fourth Anniversary Since Launch with 75.0% Market Share in Third Quarter, According to IDC*. June. Available at http://www.idc.com/getdoc.jsp?containerId=prUS23771812.

Mann, Ronald. 2006. Commercializing open-source software: Do property rights still matter? *Harvard Journal of Law and Technology* 20 (1) (Fall): 30–31.

Mansfield, Edwin, Mark Schwartz, and Samuel Wagner. 1981. Imitation costs and patents: An empirical study. *Economic Journal* (December): 270.

Morrison, P.D., J.H. Roberts, and D.F. Midgley. 2004. The nature of lead users and measurement of leading edge status. *Research Policy* 33 (2): 351–362.

Parloff, Roger. 2007. Microsoft takes on the free world. *Fortune Magazine.* Available at http://money.cnn.com/magazines/fortune/fortune_archive/2007/05/28/100033867/, p. 1.

Rao, P.M., J.A. Klein, and Erik Borg. 2006. Globalization of the ICT sector and enforcement of intellectual property rights: An open source software perspective. ITS 16th Biennial Conference, Beijing, June.

Rao, P.M., J.A. Klein, and Ramdas Chandra. 2011. Innovation without property rights and property rights without innovation: Recent development in the ICT Sector. *Advances in Competitiveness Research* 19 (No. 1/2).

Schacht, Wendy H. 2000. 98-862: R&D partnerships and intellectual property: Implications for U.S. policy. *Congressional Research Service* (December 6).

Shapiro, Carl. 1999. Exclusivity in network industries. Economic Analysis and Policy Group. University of California, Berkeley, p. 674. Available at http://faculty.haas.berkeley.edu/shapiro/exclusivity.pdf.

Teng, Tina. 2010. Android becomes smart phone market's secret sauce for success in Q2. Press Release. iSuppli Corp. (October 12). Available at http://www.isuppli.com/Mobile-and-Wireless-Communications/News/Pages/Android-Becomes-Smart-Phone-Markets-Secret-Sauce-for-Success-in-Q2.aspx.

U.S. Department of Justice (DOJ) & Federal Trade Commission (FTC) Antitrust Guidelines for the Licensing of Intellectual Property, Section 2.1 (1995). Available at http://www.justice.gov/atr/public/guidelines/0558.htm (DOJ-FTC Antitrust Guidelines).

U.S. Patent and Trademark Office. 2012. All technologies report. January 1, 1987 to December 31, 2011. Available at http://www.uspto.gov/go/stats/all_tech.pdf.

Ventre, Louis, Jr. 2011. Utility patent application costs. Available at http://www.lventre.com/details.html#others.

Von Hippel, Eric. 2005. *Democratizing innovation.* Cambridge, MA: MIT Press.

World Intellectual Property Organization. *Understanding Copyright and Related Rights (WIPO Copyright).*

World Intellectual Property Organization. *Understanding Industrial Property (WIPO Industrial Property).* Available at http://www.wipo.int/freepublications/en/intproperty/895/wipo_pub_895.pdf.

5 Inter-Firm Technology Transfers and Collaboration

Chapter Overview

This chapter discusses high-tech collaborative relationships and antitrust considerations in the context of appropriability issues. Specifically, we discuss the open innovation paradigm, inter-firm technology transfers, the cloud computing/application service model, and inter-firm technology collaborations.

Appropriability Issues Versus the Need for Collaboration

Appropriability in Perfect Competition Model

In a perfect competition setting where no producer has market power and all firms have access to the same technologies, the rate of innovation in a market economy would be expected to be very low (López 2009). The problem arises from the difficulty that a creator of new knowledge has in reaping the full return from the creator's investment (Arrow 1962). The ability of rivals to imitate at very low costs means the knowledge that one creates—often at a high sunk cost—creates a market failure because of the difference between the private and the social marginal return on any new knowledge being generated. Therefore, a market based on a perfect competition model would be expected to lead to underinvestment in R&D activities. In reality, few markets approach the perfect competition model, and none do in high-tech fields, so the risk of some degree of market failure remains.

Drivers of Appropriability

Teece (1986) has stated that appropriability regimes, ranging from tight to loose—in terms of the innovator's ability to realize greater or lesser profits from an innovation—are a function of the nature of the technology and the type and effectiveness of the available legal protection mechanism. Hence the use of various legal mechanisms to protect the innovator's intellectual property (IP), such as patents, that we examined in chapter 4.

Moreover, the development and application of technology to make products and services that customers value enough to purchase are heavily dependent on the success of high-tech companies' competitive differentiation strategies—including exploitation of lead times by ramping up early to achieve scale economies in manufacturing and marketing resources, targeting certain markets, effective implementation of policies and practices to maintain the secrecy of high-value innovative research especially before bringing the new product to market, pricing, and, branding.

The Different Appropriability Mechanisms

Firms use a mixture of means to protect their innovations, the composition of which may depend on the stage in the life cycle of the industry sector in which they operate (Dosi, Marengo, Pasquali 2006).

Based on an extensive survey of quantitative and case studies, López concluded that (1) "firms tend to employ different appropriability mechanisms. Sometimes they do it sequentially—e.g. an invention is protected by secrecy at a early stage and later on is patented—and at other times simultaneously—e.g. because an invention comprises many elements that can be protected through different appropriability tools," and (2) "lead time and secrecy seemed to be the most relevant appropriability devices for most sectors and innovation types," as opposed to legal intellectual property mechanisms such as patents (López 2009).

Nevertheless, complex technologies are often too cumbersome, inter-connected with other technologies, and expensive for any single firm to master alone. Legal IP protections and competitive differentiation strategies exercised by a firm operating alone cannot overcome these characteristics, which are inherent in complex technologies. Thus, in fast-moving markets that change continuously with new products, functions, and features appearing on a regular basis, companies are forced to innovate by acquiring access to or developing new technologies. This will necessarily involve going outside a firm's in-house research and technical organizations to access innovations happening in the same or complementary fields as the firm's. The latest data, shown in Table 5.1, suggest the rapid growth of technology alliances during the decade following 1996. For example, alliances between U.S.-owned and foreign-owned firms have grown by 48 percent. And alliances between foreign-owned firms have grown by 146 percent.

Teece explained the primary reason for this rapid growth: "Today few companies can afford to develop internally all the technologies that might provide an advantage in the future" (Teece 2000, 132). He also explained that there is a "mix of approaches" in obtaining access to externally developed technologies, such as technology purchases, licenses, partnerships, and alliances. However, as Teece also pointed out, certain critical technologies are still developed internally. Figure 5.1 illustrates this choice of organizational options as a function of the age of the technology.

Table 5.1

U.S. Industrial Technology Alliances with U.S. and Foreign-owned Companies, Worldwide, by Country/Region of Partner: 1996, 2001, and 2006

Alliance Type	1996	2001	2006
U.S. only	243	210	249
U.S.-Foreign	240	235	356
U.S.-Europe	149	151	207
U.S.-Japan	48	34	54
U.S.-Asia (excluding Japan)	21	19	50
U.S.-Canada	13	26	31
U.S.-Latin America	1	1	4
U.S.-Middle East	8	4	10
Foreign-Foreign	119	157	293

Source: National Science Board (2010).

Figure 5.1 **Age of Technology and Organizational Form**

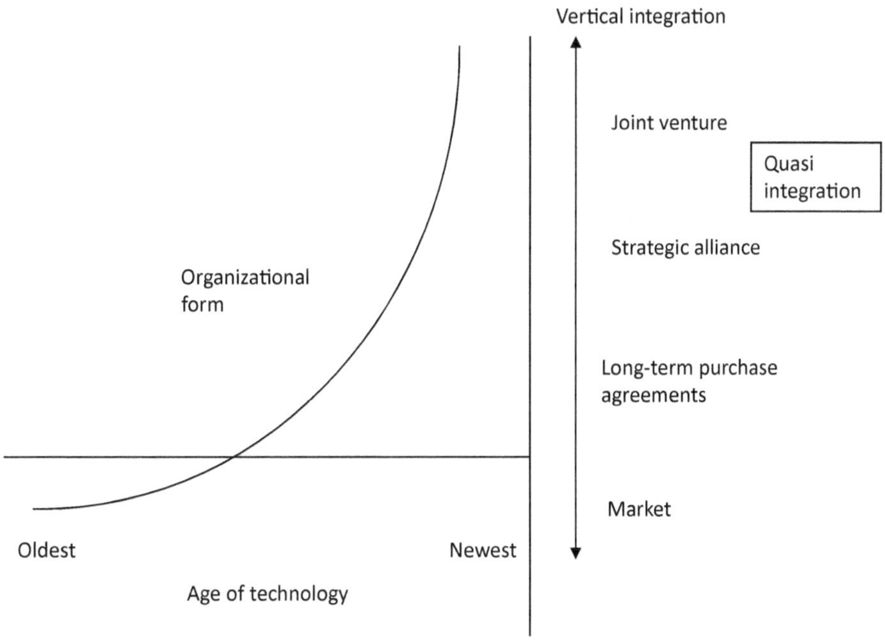

OPEN INNOVATION PARADIGM

The increasing trend toward cross-licensing, strategic technological alliances, and joint ventures may be viewed as indicators of the emerging open innovation (OI) paradigm, which has received much attention in recent years (Chesbrough 2003). In contrast to the traditional vertically integrated (VI) closed innovation model where a firm relies primarily on its own internal resources to develop and market its proprietary innovations, a company following the OI model commercializes internal as well as external ideas by utilizing external as well as in-house pathways to the market.

In the OI model, large companies are seen more and more as becoming "market coordinators of increasingly distributed and vertically disintegrated value chains" (Christensen 2005). And as Chesbrough (2003) suggests, the boundary between a firm and its surrounding environment becomes more porous, enabling innovative activities to move easily between the two. As such, OI is the antithesis of the VI model; it challenges the core competency perspective of relying exclusively or primarily on in-house corporate innovation; and it emphasizes research and production via collaboration and networks.

Firms adopting OI recognize the value of external intellectual property as well as profits to be made from others' use of its intellectual property through licensing, joint ventures, and the like. For example, in the pharmaceutical industry, Pfizer exemplifies a firm that uses external innovations in its product mix and markets internal as well as external innovations. In the case of software, the most extreme form of open innovation is manifested in the open source model discussed in chapter 4, where developers are free to access and build upon the efforts of others without paying royalties.

Open innovation necessarily depends on mechanisms to enable technology transfer, the process by which a technology, expertise, know-how, or facilities developed by one individual, enterprise, or organization is transferred to another individual, enterprise, or organization through one-way or mutual exchanges of information and/or access to the other party's technology, shared research and development, and so forth. Effective technology transfer results in the commercialization of a new product or service or in the improvement of an existing product or process.

HIGH-TECH COLLABORATIVE RELATIONSHIPS

Research on technology alliances points to motives for firms to enter into them such as market access, reducing time to market, cost, complementary technologies, and creating network effects that could be a tool for reaping profits from their innovations (Hagedoorn and Narula 1996; Gomes-Casseres 1996; Hagedoorn, van Kranenburg, and Osborn 2003).

A report by the Congressional Research Service (CRS) summarized findings by PriceWaterhouseCoopers (PwC), which stated that firms partnering with other firms:

- Invest in more new product development
- Have a lesser number of new products, although the ones that emerge are big winners
- Grow larger than others in terms of employee size and gross revenues
- Demonstrate greater productivity growth
- Experience explosive growth (Schacht 2000)

From the work undertaken by PwC's predecessor Coopers and Lybrand (1997), it appears that "collaborative growth firms are spending more on new product development while focusing more on bigger winners and on innovation . . . [and] . . . are not reluctant to go outside their organization to work with others in the development of their innovative new products."

Of the fastest growing U.S. firms, PwC found that nearly two-thirds are involved in strategic partnerships of some kind (2002).

NATURE OF A HIGH-TECH COLLABORATIVE RELATIONSHIP

There are a variety of mechanisms to effectuate technology transfers between and among unaffiliated firms. They range all the way from relatively straightforward sales of IP and one-way licenses (e.g., software licenses, patent licenses, trademark licenses) to intermediate forms of inter-firm collaboration, such as cross-licensing or mutual exchanges of technical information, to more intensive and complex forms of inter-firm collaboration, such as strategic alliances and joint ventures.

The World Intellectual Property Organization (WIPO) has identified a number of key factors in defining the nature of the relationship between and among unaffiliated parties who are considering inter-firm technology transfers (see Exhibit 5.1 for list of factors). Understanding which and how each of these factors applies in a given negotiation, and the kind of relationship that the parties wish to have with each other, is critical in determining the type of agreement that is most suitable to document the relationship.

INTER-FIRM TECHNOLOGY TRANSFER AGREEMENTS

The following is a description of forms of agreements under which technology is transferred or shared (WIPO 2004). These agreements vary in terms of strength of legal protections to prevent misappropriation of intellectual property.

Sale or Assignment of IP Rights

The first legal method of technology transfer is the sale or assignment by the owner of all his or her exclusive rights to some or all intellectual property rights embodied in a technology, and the purchase of those rights by another person or legal entity. As in the sale of a house, the title to the property moves outright from one party to another and there is normally no sharing of the use of the property except perhaps for a transitional support period or unless the new owner is willing to license back to the original owner certain usage rights. The seller or assignor of the intellectual property rights embodied in a technology has in effect monetized the value of the intellectual property being transferred as expressed either in a purchase price or through the value of the rights to other technologies received in return, but can no longer privately appropriate the intellectual property sold or assigned. However, just as selling one house does not mean that all of the owner's houses are sold or that her automobiles

Exhibit 5.1 **Key Factors in Determining the Nature of a High-Tech Collaborative Relationship**

- Stage of subject technology's development and degree of complexity
- Business objectives of parties providing and receiving the technology, their pre-existing relationships with each other, if any, and whether they are competitive with each other
- Financial, technological and marketing strength of each party and resulting negotiation leverage
- Capability/willingness of each party to commit resources to further development, implementation and marketing of products incorporating or benefiting from the technology
- Proposed terms and conditions for transfer such as:
  - Permitted scope and duration of use by the transferee,
  - Price,
  - Support offered during and after transfer in implementing and maintaining the new technology,
  - Intellectual property rights over improvements or modifications made by each party, and
  - Marketing rights and responsibilities regarding any products or services developed from the technology being transferred;
- Issues concerning allocation of risk such as:
  - Performance warranty—promise that product or service will conform to certain performance standards or specifications,
  - Non-infringement warranty—promise that the product or service will not infringe the intellectual property rights of a third party,
  - Indemnification—making a party whole for any damages awarded in a lawsuit against that party brought, for example, by the real owner of the intellectual property incorporated in the product or service that the indemnified party obtained from another,
  - Remedies and limitation of liability—limiting the parties' overall financial exposure for damages under a contract by excluding recovery of lost profits and capping the amount of direct damages that can be recouped;
- Termination rights—the circumstances under which each party is entitled to exit the agreement and any financial costs and disposition of assets associated with such termination.

Sources: World Intellectual Property Organization (2004). Note that in addition to cited sources and U.S. federal government Web sites, a major source for material in this chapter is information published by the World Intellectual Property Organization; see also *A Guide to Consortium Development and Intellectual Property Issues*, Industry Canada (March 2004).

are sold, so the sale of one type of intellectual property does not mean that any other type of intellectual property has been automatically transferred.

For example, if the sale involves a patented invention, the owner may decide to transfer all the exclusive rights to the patented invention to another person or legal entity, without any restriction in time or any other condition. However, the sale might not include the right to use the owner's trademarks in association with the patented invention. For example, a smartphone manufacturer with a patent on a particular application may license the right to other manufacturers to make and sell the patented application but not permit the licensees to use the manufacturer's distinctive logo in their products.

License Agreements

A license is the grant of permission by the owner of certain intellectual property rights, embodied in or associated with a technology, to another person or legal entity to use such intellectual property rights under specified terms and conditions. It is analogous to renting, rather than buying, a house. The owner permits the licensee (like the renter in the house example) to use his or her property but does not transfer ownership title or all rights of use.

Patents, trademarks, copyrights, and trade secrets may be licensed separately or together. Software licenses granting the licensee the right to use software owned by another party establish contract rights and obligations regarding the use of the tangible software product itself, as well as some or all of the intangible intellectual property rights that are embodied in or associated with the software.

Licenses do not guarantee that the owner of the subject intellectual property will be able to prevent misappropriation of the owner's IP. A patent license would be expected to provide the strongest legal protection, assuming the validity and strength of the underlying patent being licensed. Licenses of software that is not patented provide legal protection against misappropriation, but its efficacy depends on the nature of the software being licensed. Commoditized software that can be easily copied and requires minimal ongoing support is subject to piracy. More complex software requiring significant maintenance and enhancement support from its developer, who does not share the source code with the licensee, is less subject to misappropriability by the licensee.

Patent Licenses. A patent license permits the licensee to perform, in the country and for the duration of the patent rights, one or more of the "acts" covered by the exclusive rights to the patented invention in that country—such as the making or using of a product that includes the invention, the making of products by a process that includes the invention, or the use of the process that includes the invention. The patent license documents the conditions of use, acknowledgment of the owner's title to the patent, and payment terms (normally in the form of a royalty calculated on the basis of the number of units sold utilizing the patent or as a percentage of revenues derived from such sales).

A condition in a patent license may state that the licensee use the patented technology only for the manufacture of products in a specified field of use—for example, the manufacture of a pharmaceutical product for use by humans but not for use on animals. Alternatively, the license might specify the geographical territories in which the sale of the product embodying the patented invention may be sold.

Software Licenses. Some software and patent licenses may be (but are not always) accompanied by the transfer of know-how, which can be communicated in tangible form if sufficiently codified (e.g., design documents, software user documentation, drawings or blueprints of machines used in the patented invention, instructional manual, etc.). This form of communication requires little ongoing inter-firm day-to-day interaction, which is more characteristic of collaborations, except to answer questions. Technical support for software is often purchased as a separate service, with the software vendor assuming responsibility for ongoing maintenance and enhancements while holding back significant information that would enable the licensee to replicate the underlying software code and thereby misappropriate the vendor's IP.

Inter-firm day-to-day interactions increase when the technology being transferred is not sufficiently codifiable to be understood by the recipient without, for example, detailed explanations, training, and assistance from an engineer of the technology transferor to an engineer of the technology recipient. This relationship is still, however, closer in kind to the vertical supplier–customer model than a horizontal collaborative arrangement in which both (or multiple) parties are contributing technology and know-how to a combined initiative of some sort.

The risk that the technology and know-how being communicated by the transferor to the recipient might be improperly used by the recipient or disclosed to third persons is a very real concern that must be addressed in the license agreement itself or in a separate confidentiality agreement. In addition to defining how the technology and know-how are to be used by the recipient and to whom they may be disclosed, the agreement should specify effective monetary and injunctive remedies in order to deter a breach. See Exhibit 5.2 for examples of typical confidentiality terms in a confidentiality agreement or software/know-how license.

Exhibit 5.2 **Selected Confidentiality Terms**

- **Definition of "Confidential Information" [Defines what is considered Confidential Information that is to be contractually protected]**
  - **Example:** all information, in any form, furnished or made available directly or indirectly by one party to the other which is marked confidential, restricted, proprietary, or with a similar designation including, but not limited to, that related to the business, systems, operations, strategic plans, clients, pricing, methods, processes, financial data, programs, and/or products of the other party.
- **Exclusions from definition**
  - **Example:** "Confidential Information" shall not include information that (i) is already rightfully known to a party at the time it is obtained from the other party, free from any obligation to keep such information confidential or is or becomes publicly known or available through no wrongful act of a party; or (ii) is independently developed by a party without the use of any proprietary, non-public information provided by the other party under this Agreement.
- **Specify purpose of information exchange**
- **State limits on use and disclosure rights with respect to Confidential Information on a need-to-know basis and to meet judicial/regulatory demands**
- **Remedies for breach**
  - **Example:** the right to get an injunction to stop unauthorized use or disclosure of the Confidential Information

ALTERNATIVES TO TECHNOLOGY TRANSFER AGREEMENTS

Cloud Computing/Application Service Model

One way to get around technology transfer concerns that software application providers are increasingly using is to convert to an all-service model, with a couple of variants referred to as "cloud computing" or "application service." With these services, the customer does not obtain physical possession of the software for installation on its own infrastructure. Rather, the customer can rent the functionality of a computing infrastructure and/or application software, which is accessed online via the Internet (but with a secure connection), an Internet-based variant of what in the past has been referred to as a time share. Alternatively, the service provider runs the software application itself and provides a service to the customer that the application enables, a current Internet-based variant of what in the past has been referred to as a service bureau.

The advantage of the cloud computing service model to the customer is potentially significant savings in infrastructure costs, as the customer does not have to worry about hosting and maintaining the software itself. However, the customer is ceding control over proprietary data, which must be transferred from an internal organization to the service provider's computing environment for processing, where the data could be more vulnerable to being accessible by other users of the provider's services.

The advantage of the cloud computing service model to the service provider is that the provider maintains more control over its own high-tech assets, including its proprietary software, because it is not providing copies for the customer to use on its own computing environment. As explained by the National Institute of Standards and Technology, the service provider also benefits from economies of scale by offering its customers "network access to a shared pool of configurable computing resources (e.g., networks, servers, storage, applications, and services) that can be rapidly provisioned and released" (Jansen and Grance 2011). Finally, as a day-to-day software service provider, a high-tech firm can gain a deeper insight into the customer's ongoing needs than is possible through one-shot license and maintenance arrangements where there is less interaction with the customer. This permits more effective cross-selling.

Trademark Licenses

Trademark licenses may involve more inter-firm collaboration than other one-way intellectual property licenses, often as part of a business arrangement whereby the reputation, technical information, and expertise of one party are combined with the investment and customer relationships of another party for the purpose of marketing products or services. The license of such a mark or name by its owner is normally combined with the supply by that owner of know-how in some form, such as technical information, technical services, or collaboration in production, marketing, and customer support. The risk to the trademark owner is the impact of the licensee's actions on the quality of the owner's brand. See Exhibit 5.3 for examples of some typical trademark license terms.

Cross-Licensing

Cross-licensing of patents is one means used by high-technology firms to allow each party to use the other party's patents included within the cross-licensing arrangement without fear of a patent infringement suit brought by the patent owner (assuming that each licensor has the rights to license the patent to the licensee). There is little in the way of ongoing collaboration other than access to the right to use an expanded or complementary set of technology assets.

INTER-FIRM TECHNOLOGY COLLABORATIONS

Inter-firm technology collaborations increase the access to each firm's technologies and related technical information and expertise, which can help reduce costs of go-it-alone R&D. They bring together complementary assets and skills, and permit work to be done that is too expensive for one company to fund. They can fill product line

Exhibit 5.3 **Examples of Trademark License Terms**

- State the purpose of trademark license—for example, promoting and marketing the customer solution created as a result of the parties' integration of their separate products and promoting and marketing the related products and services of each party
- Retention of brand ownership conferred by trademark—limited trademark license granted to licensee for stated purpose with no right to transfer or allow use by third parties
- Brand quality control—Assurances that trademark owner's instructions on use of its trademarks are strictly followed and right of trademark owner to review materials using its trademark in advance for its approval
- Avoid brand confusion—Parties agree not to register or use any mark that is confusingly similar to any of the other party's trademarks

gaps. And most importantly for the marketing strategy of high-tech firms, they can speed time-to-market of products or services incorporating the elements of the firms' combined technologies.

But at the same time, unaffiliated parties' economic interests, business objectives, and strategies are not normally as aligned as those of affiliated firms. Thus, structuring inter-firm collaborations to preserve the incentives of both parties to continue their cooperation can be challenging. As the World Intellectual Property Organization (2004) put it, the "secret to the success of technology transfer agreements is that each party perceives the other as a partner in a fruitful collaborative process."

In terms of optimizing high-tech firms' product marketing strategies, the more successful full-scale collaborations often (1) encompass major customer segments with a strong customer and/or installed product base that neither firm possesses alone, (2) involve synergies among the collaborators' separate technologies or products components that enable them to create a customer solution meeting specific market needs, and (3) complement (as opposed to conflict with) existing sales channels (Bleeke and Ernst 1993).

First and foremost is determining the scope of the inter-firm collaboration and the stage or stages in which there is to be primary reliance on collaboration or primary reliance on independent action by each firm. Large high-tech firms in particular, such as IBM, use a mixture of collaborative relationships, depending on their strategic objectives. According to Bleeke and Ernst, IBM uses collaborations to "strengthen specific elements of its business system." See Exhibit 5.4 for examples of past IBM collaborations and their purpose.

Upstream Research Collaboration

Companies may decide on limited collaboration—working together to perform research, but then exploiting the results of this research within their individual organizations. The collaborating firms, sharing upstream costs and risks of the research itself, which may or may not produce anything useful, must agree up front on how they will allocate ownership of the jointly developed intellectual property or whether they will all own it without having to account to each other with respect to revenues received from downstream commercial exploitation.

Exhibit 5.4 **IBM Collaboration Examples**

1. Collaborating Party—Ferranti IBM Business Objective: Penetrate markets for PS/2 operating system through inclusion of IBM PCs in Ferranti computer	*3. Collaborating Party—Microsoft* IBM Business Objective: Joint software development enabling IBM to improve its competitive market position in the PC business
2. Collaborating Party—Siemens IBM Business Objective: Share development costs of advanced DRAM	*4. Collaborating Party—Toshiba* IBM Business Objective: Joint development of LCDs enabling IBM access to Toshiba technology

Source: Bleeke and Ernst (1993).

Strategic Alliances and Teaming Among High-Tech Firms

Collaboration can also take the form of strategic alliances among firms with complementary technology assets and/or marketing assets such as customer contacts. One variant is where two parties combine each of their separate products into an integrated solution to meet a defined customer need—for example, a software application that displays data in an easily usable fashion and data products that can be loaded into and displayed on the application. These collaborations can vary in length and complexity. Where customization of each product is involved to enable an integrated customer solution, the ownership of the customizations can be an issue that often needs to be negotiated along with control over the customer relationship and limits on each collaborator's ability to offer competing products and services using the combined solution outside the scope of the collaboration.

Collaborations Among High-Tech Firms and Their End-Users

Alternatively, the collaborative process can involve cooperative innovation among high-tech firms and their end-users, in response to unmet market needs. The emphasis is on enabling the rapid availability and diffusion in the marketplace of the technology in products or processes that are readily usable by customers. Early adopters of the product can potentially gain a time-to-market advantage over their competitors and provide useful input to the technology firm that can help influence the direction that the product's technology takes. The technology firm can use the collaboration to establish and reinforce a long-term relationship with its customers who participate, and to obtain intelligence on market needs that it can use to further develop its innovations and provide more useful products.

Joint Ventures

Joint ventures cover a broad range of inter-firm collaborations that may be contractual in nature or embodied in the form of a separate legal entity owned by the joint venturers, such as a partnership or corporation. Aside from upstream research ventures mentioned earlier, joint ventures can focus on product development, production, or marketing/sales, or can encompass all of these elements in a core business joint venture.

Setting realistic expectations about business objectives and anticipated economic value of the joint venture during the course of negotiations is essential to success. Scope, size, and duration of the joint venture need to be addressed upfront. There are tradeoffs that must be considered. More resources devoted to the joint venture can increase its potential market footprint, but this may be at the expense of the independence and profit maximization of each participating firm in its own markets.

The most difficult issues to resolve when putting together a successful high-tech joint venture include the following:

Governance. How are significant decisions made? Does each joint venturer have veto power over strategic matters such as changing the joint venture's scope and business objectives, budgets, and taking on new partners? If there is a separate legal entity with a board or operating committee, is a unanimous, super-majority or plain majority vote sufficient, and for which type of decisions? If the legal entity has employees who are managing the day-to-day operations of the joint venture, how much discretion should be delegated to them? What mechanisms need to be in place to balance flexibility to meet changing market circumstances versus sufficient controls to ensure that the joint venture continues to meet the owners' collective needs?

Funding. How are initial funding and ongoing budget contributions allocated? Will the contributions be equal or apportioned according to some size-based formula? If one or more joint venturers contribute technology rather than money, how will that technology be valued and what rights will the joint venture get to the contributed technology versus rights retained individually by the contributor(s)? If there are multiple projects, some of which may involve upstream research that is core to supporting a variety of other more downstream projects, to what extent should core funding be mandatory versus discretionary in order to avoid free-rider problems? How are resources prioritized in the event of multiple projects?

Marketing. To what extent will marketing of jointly developed products be joint, a mixture of joint and individual, or purely individual? Will the joint venture products be separately branded or incorporated in the owners' own brand portfolios? Will there be a lead joint venture principal? If the legal entity has employees, will there be a manager or group in charge of the joint venture's marketing and sales? How will the sales revenue be allocated? Will some of the revenue be plowed back into more joint venture activities or split after covering expenses among the joint venturers equally or according to some size-based formula?

IP Ownership. Who will own the intellectual property developed by the joint venture? Will it be owned jointly by the joint venturers? Will the contributor of the base technology own the derivative technology developed from it and grant its co-venturers and the joint venture itself (if a separate legal entity) a license to use and distribute?

How broad will that license be? If the joint venture is a separate legal entity, would the joint venture own the IP developed by the joint venture with licenses to each of the joint venture owners?

Competition. Will the joint venture owners be able to compete with their own products against the products created by the joint venture and vice versa? To what extent can the technological fruits of the joint venture be incorporated in each joint venturer's competing products? How inclusive will the joint venture be in terms of allowing other market participants to be added to the original group of joint venturers?

Exit. What is the mechanism for dissolving the joint venture? How are the assets, including any joint venture–owned IP and monies, distributed to the joint venturers? Would it make more sense, if the joint venture is a separate legal entity, to realize the cash value of the joint venture by selling the joint venture entity or its assets either to one of its owners or to a third party?

ANTITRUST CONSIDERATIONS

The competition element is influenced not only by the business consideration of the co-venturers but also by antitrust laws. Collaborations among competitors raise antitrust concerns if they are used primarily as vehicles to restrain competition among the collaborating firms, to exchange competitively sensitive information, or to engage in a group boycott of competing firms not included in the joint venture. If challenged, the collaborating competitors would have to prove that (1) their joint venture or other form of collaboration has an efficiency-enhancing goal that increases competition by reducing price, raising output, increasing quality or innovation, or meeting a customer need that the competitors can only provide jointly, (2) the joint activities are necessary to achieve the efficiency-enhancing goal with any restrictions on each competitor being no more than needed for that purpose, and (3) they have safeguards in place to protect against spillover effects (i.e., exchanging price, output, or competitive marketing information about the markets in which the parties compete outside the business that is the subject of the joint venture or other collaborative activities).

SUMMARY

In this chapter, we have examined inter-firm collaborations in high-tech industries, which have arisen because complex technologies are often too cumbersome, inter-connected with other technologies, and expensive for any single firm to master alone.

Rather than try to rely primarily on its own internal resources to develop and market its proprietary innovations, a high-tech firm willing to embrace collaborations with other unaffiliated firms is following the open innovation (OI) paradigm in contrast to the traditional vertically integrated (VI) closed innovation model. A company following the OI model emphasizes research and production via collaboration and networks, and commercializes internal as well as external ideas by utilizing outside as well as in-house pathways to the market.

We reviewed the various approaches that are used by firms to obtain access to externally developed technologies, which range all the way from relatively straightforward sales of intellectual property and one-way licenses (e.g., software licenses, patent licenses, trademark licenses) to intermediate forms of inter-firm collaboration, such as cross-licensing or mutual exchanges of technical information, to more intensive and complex forms of inter-firm collaboration, such as strategic alliances and joint ventures.

The more complex forms of inter-firm technology collaboration increase participating firms' access to each other's technologies and related technical information and expertise. They bring together complementary assets and skills, and permit work to be done that is too expensive for one company to fund. They can fill product line gaps. And, most importantly for the marketing strategy of high-tech firms, they can speed time-to-market of products or services incorporating the elements of the firms' combined technologies.

But at the same time, as we noted, unaffiliated parties' economic interests, business objectives, and strategies are not normally as aligned as those of affiliated firms. Thus, structuring inter-firm collaborations to preserve the incentives of both parties to continue their cooperation presents significant challenges, which we have described along with specific factors to consider in trying to successfully address those challenges.

KEY TERMS

appropriability
cloud computing/application service model
joint ventures
open innovation (OI) paradigm
technology transfer agreements
upstream research collaboration

DISCUSSION QUESTIONS

1. Why would a market based on a perfect competition model be expected to lead to underinvestment in research and development activities?
2. What are the characteristics of complex technologies that make it difficult for any single firm to master alone?
3. Describe the open innovation model and apply the model to an example in a high-tech industry.
4. Explain the reasons why high-tech firms enter into collaborative relationships with other high-tech firms and identify key factors in determining the nature of the high-tech collaborative arrangements they decide to enter into.
5. What benefits and risks are there to the owner of software who licenses the software to a third party? What means can the software owner use to reduce his or her risk?
6. What kinds of business issues need to be resolved during negotiation of a high-tech joint venture in order to increase its likelihood for success?

REFERENCES

Arrow, Kenneth. 1962. Economic welfare and the allocation of resources for invention. In *The Rate and Direction of Inventive Activity: Economic and Social Factors*, ed. Richard R. Nelson, N.B.E.R. Special Conference Series, vol. 13, Princeton, NJ: Princeton University Press.

Bleeke, Joel, and David Ernst, eds. 1993. *Collaborating to compete: Using strategic alliances and acquisitions in the global marketplace.* New York: John Wiley & Sons.

Chesbrough, H. 2003. *Open innovation: The new imperative for creating and profiting from technology.* Cambridge, MA: Harvard Business Review Press.

Christensen, Jens Frosley. 2005. Withering core competency for the large corporation in an open innovation world? Druid Tenth Anniversary Summer Conference, Copenhagen Business School, Copenhagen, Denmark.

Coopers and Lybrand L.L.P. 1997. Partnerships pay off for growth companies, Trend Setter Barometer. Available at http://www/barometersurveys.com.

Dosi, G., L. Marengo, and C. Pasquali. 2006. How much should society fuel the greed of innovators? On the relations between appropriability, opportunities and rates of innovation. *Research Policy* 35: 1110–1121.

Gomes-Casseres, Benjamin. 1996. *The alliance revolution: The new shape of business rivalry.* Cambridge, MA: Harvard University Press.

Hagedoorn J., and R. Narula. 1996. Choosing organizational nodes of strategic technology partnering and sectoral differences. *Journal of International Business Studies* 27 (1): 265–284.

Hagedoorn, J., H. van Kranenburg, and R.N. Osborn. 2003. Joint patenting amongst companies—Exploring the effects of inter-firm R&D partnering and experience. *Managerial and Decisions Economics* 24 (2–3): 71–84.

Jansen, Wayne, and Timothy Grance. 2011. Guidelines on security and privacy in public cloud computing. Draft Special Publication 800–144. National Institute of Standards and Technology (NIST), January.

López, Andrés. 2009. Innovation and appropriability, empirical evidence and research agenda. *The Economics of Intellectual Property.* World Intellectual Property Organization, January. Available at http://www.wipo.int/export/sites/www/ip-development/en/economics/pdf/wo_1012_e_ch_1.pdf.

National Science Board. 2010. www.nsf.gov/statistics/seind10/pdf/c04.pdf.

PriceWaterhouseCoopers. 2000. Partnerships have big payoffs for fast-growth companies, Trendsetter Barometer, August 26, 2002. Available at http://www.barometersurveys.com.

Schacht, Wendy H. 2000. R&D partnerships and intellectual property: Implications for U.S. policy. Congressional Research Service (December 6): 98–862.

Teece, D. 1986. Profiting from technological innovation: Implications for integration, collaboration, licensing, and public policy. *Research Policy* 15: 285–305.

———. 2000. *Managing intellectual capital: Organizational, strategic, and policy dimensions.* New York: Oxford University Press.

World Intellectual Property Organization. 2004. Technology transfer agreements: Overview of contractual terms (May–June). Available at http://www.techmonitor.net/tm/images/9/9e/04may_jun_sf5.pdf.

6 Product and Promotion Strategies in High-Tech Firms

Chapter Overview

Product strategy is central to creating and sustaining competitive advantage in high-tech and low-tech firms alike. Our objective in this chapter is to address product strategy issues that are particularly important to high-tech firms. They include co-alignment of technology-based product concept with segmentation, issues concerning design and standards, diffusion of innovations, branding innovations, marketing of technologies, open innovation, and platform strategy.

Product Strategy

The standard textbook treatment of product strategy involves several key elements, including statement of objectives in terms of growth and profitability; strategic alternatives to achieve them (penetration of existing market versus developing new markets and products, for example, as shown in Figure 2.2); selection of customer and competitor targets; positioning strategy (e.g., cost versus differentiation strategies); branding strategy; and product strategy over the life cycle. The reader is asked to consult Lehmann and Winer (2005) for a comprehensive discussion of these areas, which apply to all products, high-tech or otherwise—especially consumer products.

Market Testing

In marketing, not much really happens until someone comes up with an idea for a product that is aimed at an unmet need of a market segment, or segments, and then takes the idea at least through successful concept testing. This is followed by *product testing*, that is, physically producing the product and getting small samples of consumers to use it. Once product testing proves successful, real-life market tests are conducted with other elements of a marketing program—promotion, pricing, and distribution—in full play.

SEGMENTATION

Technology-based Product Concept and Segmentation

It is often said that what matters to consumers is the benefit that technology offers rather than the technology itself. This holds even when technology and product are one and the same, as will be discussed later in the chapter. Still, in high-tech markets, there is a greater need for co-alignment of product concepts to technology than is generally recognized in the marketing literature for non-high-tech markets. For example, in the total product concept, in terms of the customer benefit hierarchy shown in Figure 6.1, technology plays a central differentiating role at each level of benefits. Therefore, the benefits are aligned with the technology. This concept was first developed by Levitt (1980). The reader is also referred to Lehmann and Winer (2005).

The product concepts include:

- *Generic Product.* This is one that offers basic benefits—functionality without frills—and competes on the basis of price. For example, a generic laptop computer could be described in terms of memory capacity, weight, screen size, width, the number of USB ports, processing speed, and its graphics card.
- *Expected Product.* This is one whose additional benefits customers have come to expect. This could mean newest operating system, wireless Internet access, enhanced battery power, Intel processor, large hard-drive capacity, and a warranty.
- *Augmented Product.* This offers additional features and benefits beyond expectations. For a laptop these features could include being unusually lightweight, having additional security features such as a finger scan, or having a high-definition screen or Blue Ray player, for example.
- *Potential Product.* This concept, on the other hand, is one whose features and benefits can be added in the future such as memory capacity, upgrades of the operating system, a better battery, and occasionally the ability to update the graphics card.

The point is the bundle of features and benefits at each level involves technology as the underlying source of differentiation. Indeed, without the underlying technologies, neither the laptop that is of much value, nor the benefits it brings, would exist. The augmented and potential product concepts correspond to high levels of differentiation and enable firms to focus on nonprice competition. Note that over time, as a product reaches maturity stage and newer and better products are introduced in the same category—often at the same or an even lower price—consumer expectations about an augmented product rise, and today's augmented product becomes tomorrow's expected product.

These concepts can be easily co-aligned with segmentation, targeting, and the positioning strategies of firms. Let us say, as shown in Figure 6.1, that the market for laptops is divided into four segments of consumers based on the four product concepts noted above and the respective benefits associated with each. The four segments include Basic,

Figure 6.1 **Total Product Concept and Its Relationship to Segmentation: The Case of the Laptop Computer**

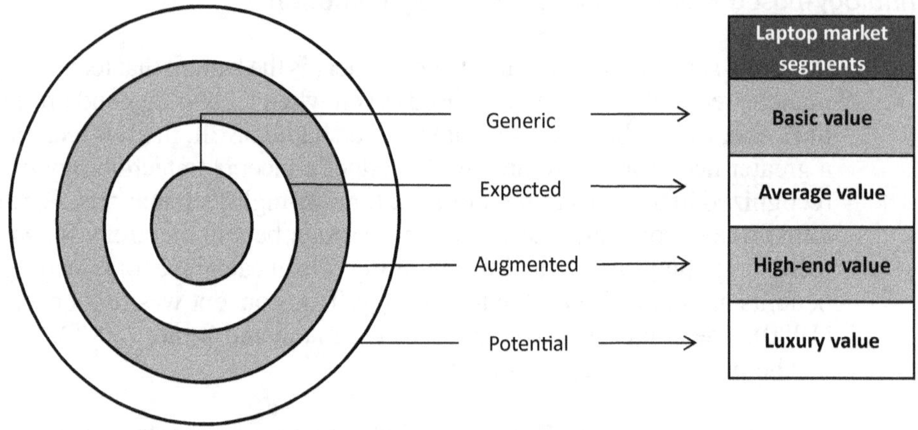

Source: Adapted from Levitt (1980) and Lehmann and Winer (2005).

Mid-Range, High-End, and Luxury. A producer of laptops may choose to target one or more of these segments, developing an appropriate marketing mix and positioning strategy that communicates the benefits of the product. Consider how starting with the definition of the market as the "market for laptops," is grounded as much in technology (e.g., wireless) as on the myriad benefits the technology provides (ability to have a mobile office). Similarly, segmenting the market for laptops—dividing the market into groups of consumers with similar needs—based on the product concepts discussed above, is based as much on features grounded in technology as in the benefits of convenience such features offer. Indeed, in high-tech markets, the line between technology-related product features and the benefits derived from such features is often indistinguishable. Likewise, selection of a target market(s) in which to compete depends as much on the depth and breadth of a firm's technological capability as on other competitive factors, such as the discovery of an unserved or underserved segment. Finally, technology-related product features make just as effective positioning statements as those that emphasize the benefits such features bring to the consumer.

Disruptive Technologies and Segmentation

Clayton Christensen (1997) argues that a "disruptive technology"—a new technology having lower cost and performance measured by traditional criteria, but typically simpler, smaller, and more convenient to use—may enter and expand emerging market niches, improving with time and eventually threatening established products in their traditional markets. In a subsequent version of the concept, Christensen and Raynor (2003) make a distinction between *low-end disruptions*, which address the low end of an existing value network, and *new-market disruptions*, which create new value networks. An example of low-end disruptive technology is the 5.25-inch floppy disk drive, which disrupted the market for the vastly superior 8-inch disk drive, which was not affordable for desktop computers. The market for 5.25-disk drive was later disrupted by 3.5-inch drive, the market for which was in turn disrupted by CDs and USB flash drives. Low-end disruptions start at the low-end segment of an existing market by offering low-cost models to established firms' least profitable customers. An example of new-market disruption is Canon's desktop photocopier, which expanded the market for photocopying by making photocopying in one's office easy and convenient (Christensen and Raynor 2003). In contrast to low-end disruptions, new-market disruptions *create* a need in customers and hence an entirely new segment.

Indeed, as Utterback and Acee (2005) suggest, the true importance of disruptive technology lies in its ability to expand and broaden the market and provide new functionality rather than merely displacing incumbent firms' established products. Christensen distinguishes disruptive technologies from what he calls sustaining technologies that are common to large companies and involve improving a product in an established market. Large companies typically ignore disruptive technologies until their profitability becomes noticeable. Although many of them enter the market to develop products based on disruptive technologies, few succeed. This is because they follow an unwinnable dual strategy of continued resource commitments to old products and technologies while simultaneously pursuing products based on disruptive technologies. Allocation of resources to old technologies continues because many established customers prefer to have improved versions of existing products rather than products based on completely new technologies.

Moreover, as Cooper and Schendel (1992) suggest, decisions concerning allocation of resources between old and new technologies are influenced by the perceived threat to old skills and positions as well. As a result, many large companies with older technologies are left behind. Meanwhile, companies with products based on disruptive technologies enter and expand the emerging market segment to early adopters and beyond and begin to compete with the established products in their traditional markets. To be sure, a disruptive technology is not to be confused with a radical innovation such as the transistor, which simply replaced the tube (Utterback 1994). From our standpoint, appropriability of established innovations is threatened regardless of whether new technologies are disruptive or radical.

DESIGN ISSUES IN PRODUCT STRATEGY

Firms' strategies affect and are affected by the presence or lack of a dominant design that is widely accepted by the marketplace. A firm with a dominant design can lock out its competitors and stage future generations of products in the category. On the other hand, uncertainty surrounding the lack of dominant design or its late emergence can result in potential customers delaying their purchases of the product, which slows the market growth.

Shapiro and Varian (1999) note several factors that contribute to gaining first-mover advantages: being a pioneer in innovation, rather than a follower; building confidence among consumers; being suppliers of complements and distributors; advocating education and the sharing of information; appropriate pricing and promotional strategies; and exploiting experience curve effects.

In a significant empirical study on the emergence of dominant designs, Srinivasan, Lilien, and Rangaswamy (2006), by utilizing data on sixty-three product categories, developed and tested a model for the relevance of five product characteristics to the emergence of dominant design and its timing. The characteristics include *appropriability*, *network effects*, the network of firms in the *value net*, the type of *standards-setting process* (de jure versus de facto), and *R&D intensity*. The results supported the following hypotheses:

- The likelihood of the emergence of dominant design, as well as its quick emergence, increases with weaker appropriability faced by the product category (see Figure 1.2 in chapter 1). This is because a weak appropriability regime, where firms have little or no control over prices, encourages knowledge sharing among firms and provides incentives for rapid development of complementary goods (e.g., PC and DVD).
- The emergence of a dominant design occurs within a shorter time in product categories where the value net is large. That is, firms' linkages with the suppliers and producers of complementary goods are extensive (e.g., DVD producers, film studios, and the like). Each firm in the value net will benefit from rapid emergence of a dominant design, notwithstanding greater transaction and coordination costs associated with more firms.
- Elapsed time to the emergence of a dominant design in a product category is shorter in a market-driven standards-setting process. This is because in a formal de jure process, which typically involves many players including international bodies, it is more difficult to achieve consensus. For example, the standard for the fax machine, which was set through a de jure process, took twenty-three years to emerge.
- The emergence of a dominant design in a product category involving radical technology (e.g., microwave ovens, a radical technology when they were introduced and still nothing has replaced them in the marketplace) is less likely and it takes longer when it does emerge. This is because the emergence of a radical product category evolves over several years and is fraught with high uncertainty and limited consumer acceptance of early versions.
- The greater the R&D intensity of the product category, the greater the likelihood that a dominant design will emerge. One reason for this is that product categories with high R&D intensity are more likely to produce rival designs, thus increasing the pressure for the emergence of a dominant design.

Surprisingly, the authors found no support, either way, for the impact of network effects on the emergence or timing of dominant design. Where there are network effects, consumers demand compatibility, and hence a single standard, because of the additional value each consumer derives from it.

Dominant Design/Standards

Product categories vary with respect to the emergence of a dominant design and its timing. For example, although the fax machine was introduced in 1960, the dominant design did not emerge until 1983. In contrast, the dominant design for DVD players emerged only three years after its introduction in 1996.

Some authors use the terms "dominant design" and "standards" interchangeably (Shapiro and Varian 1999), while others make a distinction between the two. Srinivasan, Lilien, and Rangaswamy (2006) suggest that the term "standards" refers to functional aspects such as quality, compatibility, adaptability, and connectivity that are required for the proper functioning of products. These standards, they argue, exist regardless of whether or not there is market acceptance, while market acceptance is critical to the emergence of dominant design. However, this distinction between standards and dominant designs becomes moot in markets with strong network effects (e.g., fax machines and modems) where compatibility is central to market acceptance.

Competition in high-tech industries often revolves around technical standards involving compatibility. Although arguments on technical standards that interlink products made by required suppliers are common (e.g., software standards to interlink Apple computers and IBM compatibles), each firm seeks to have its own proprietary technology become the industry standard. In short, there are two forces at work here. Where there are significant network externalities (e.g., telecommunications), firms have strong incentives to agree to common technical compatibility standards in the belief that everyone in the industry gains. On the other hand, firms—especially those that believe themselves to be at the cutting edge of innovation—also have an incentive to produce an incompatible product in the belief that their product could become the industry standard, thus gaining the first-mover advantages and market power.

Critical Assets Needed for the Battle over Standards. Shapiro and Varian (1999) document a fascinating set of historical examples ranging from railroad gauges to power systems to color televisions that illustrate important implications for strategy concerning standards. They identify seven critical assets a firm must possess to prevail in the battle over standards.

1. First-mover advantages
2. Control over installed base
3. Strong intellectual property rights (IPRs)
4. Low-cost manufacturing capacity
5. Continuous innovation
6. Strength in complements
7. Reputation and brand name

Controlling the Installed Base

Once a firm gains first-mover advantage, it has the ability to build a large installed base rather rapidly. This is particularly true of products and services, such as telecommunications and software, which are subject to network externalities—meaning that benefits to consumers increase as the user-set expands. However, this does not occur automatically even in the presence of network externalities. Pricing strategy—often penetration pricing or low initial price—plays a key role in building a large installed base rapidly.

Central to control over an installed base is continuous innovation, which keeps potential competitors at bay and gives users an incentive to stay with the first mover.

Strong IPRs in the form of patents and copyrights are important in controlling the installed base, as are products that offer backward and forward compatibility, scalability (ability to expand capacity), upgradability (ability to substitute old components with new and improved components), and easy servicing. Pricing strategies also play an important role in controlling the installed base. As noted in chapter 3, it is no accident that, according to one estimate, the monopoly price of Windows should be as high as 16 times the price actually charged by Microsoft. Low-cost advantages in manufacturing and strength in complementary products provide additional ammunition to the pioneering firm to create and control a large installed base. Reputation, a "company-centric" concept that focuses on credibility of the firm among a broad set of stakeholders, and brand name, a "customer-centric" concept that focuses on benefits of a product or service to the customer, also provide this ammunition (Etterson and Knowles 2008). They play a central role in convincing customers, complementors, suppliers, and distributors alike that your product will prevail in the battle over standards.

DIFFUSION OF INNOVATIONS

THE ROGERS MODEL

Diffusion is the process by which individuals and firms adopt a new/improved product, service, or technology. The literature on diffusion of innovations has a long history, with origins in sociology and economics. The data presented in Table 6.1 show that the rate at which products are diffused varies widely across product categories. For example, it only took seven years for the Internet to spread to a quarter of the U.S population compared to thirty years for the microwave oven.

The marketing literature on diffusion has been strongly influenced by Everett Rogers's sociological perspective, which culminated in several editions of his classic book *Diffusion of Innovations* first published in 1962 (see also fourth and fifth editions published in 1995 and 2003, respectively) Rogers identified five characteristics that influence the rate at which innovations diffuse and are adopted at the individual level. He considered cell phones to have an "almost ideal set of perceived attributes" (fifth edition, 2003, p. 245) for rapid rate of adoption in the United States.

1. *The Relative Advantage.* This is the degree to which users perceive an innovation as superior to the one it supersedes in economic as well as noneconomic benefits. A cell phone saves time and serves as a status symbol under certain circumstances.
2. *Compatibility.* This is the degree to which an innovation is compatible with the values, past experiences, and needs of potential adopters. A cell phone user can talk to anyone on another mobile device or someone with a traditional landline telephone.
3. *Complexity.* This has to do with the degree to which an innovation is perceived as difficult to understand and use. A cell phone is simple and easy to use and

Table 6.1

Spread of Products to a Quarter of the Population

Product	Year Invented	Years to Spread
Electricity	1873	46
Telephone	1876	35
Automobile	1886	55
Airplane	1903	64
Radio	1906	22
Television	1926	26
VCR	1952	34
Microwave Oven	1953	30
PC	1975	16
Cellular Phone	1983	13
Internet	1991	7

Source: Federal Reserve Bank of Dallas (1996).

hence has greater likelihood of rapid adoption. When an item is perceived to be complex, the rate of adoption is slower.

4. *Trialability.* This is the degree to which an innovation can be tested. A cell phone can be tried easily prior to purchase, which in turn reduces the uncertainty associated with the purchase, and post-purchase regret is minimized with thirty-day trial periods.
5. *Observability.* This is the degree to which results of an innovation are easily observed and communicated to others. The easier it is to evaluate the results of an innovation after trial, the lower the uncertainty associated with the innovation and therefore the greater the likelihood of its adoption.

As Hall (2004) notes, the above characteristics—derived primarily from a sociological perspective—are quite consistent with Hall's economic perspectives on innovation. Relative advantage can be viewed in terms of the relative profitability of an innovation, which was emphasized by Griliches (1957) in his seminal empirical study of the diffusion of hybrid corn seed from an economic perspective. Trialability and observability help mitigate uncertainty associated with any new technology, thus increasing the likelihood of its adoptability. Compatibility is an economic issue involving standards and design, and both in turn are intertwined with uncertainty and relative benefits of the new technology. Similarly, complexity of an innovation could entail considerable direct and indirect cost to the potential adopter in terms of time and paid tutoring. Note that the characteristics of innovations that are conducive to rapid rate of adoption also help provide first-mover advantages to the firms, thus increasing the appropriability of their innovative effort.

Adopter Categories

Having identified the characteristics of innovations that influence individuals' decisions to adopt, Rogers (1995 and 2003) provided a methodology to categorize adopters on the basis of how quickly a group adopts the innovations. On this basis Rogers identified five categories: *innovators*, *early adopters*, *early majority*, *late majority*, and *laggards*. Based on how individuals learn about new ideas and their aversion to uncertainty, Rogers argued that the adoption curve should have a normal distribution with two basic parameters—mean ($\bar{x}$) and standard deviation (σ), as shown in Figure 6.2 (see Figure 7.3, p. 281, in Rogers 2003). Note that this means that when cumulative number of adopters is plotted, the result is an S-shaped curve, shown in Figure 6.3.

Dominant Characteristics

In order to provide further conceptual and empirical work on diffusion, Rogers provided a description of the dominant characteristics of each adopter category. For example, the innovators are venturesome, financially secure, and are risk takers. The early adopters are opinion leaders who are respected by peers and therefore exercise significant influence on potential adopters. The early majority is deliberate in its decisions but does not lead. The late majority is a skeptical bunch who possess meager financial resources and need peer pressure to adopt an innovation. Finally, the laggards, whose resources are limited, look for certainty before they adopt.

Rate-of-Adoption Influencers

In addition to the above five attributes, which influence the diffusion process at the individual level, Rogers (1995) also identified four "external" variables that influence the rate of adoption. First, the type of innovation decision matters—that is, whether the decision is made by an individual, by a community, or by a single authority. The rate of adoption is expected to be faster when an individual makes the decision than when decisions are made by more than one person or collectively. Second, the type and choice of communication channels used to inform potential adopters about an innovation have an influence on the rate of adoption. Interpersonal communication is more important for innovators and early adopters, while mass communication is more important for late majority and laggards. The choice between interpersonal and mass media also depends on whether an innovation is complex (interpersonal is more effective) or simple (mass media is more effective). Third, the nature of the social system—meaning its norms and the degree to which users are interconnected in the community—influences the rate of adoption. The greater the degree to which users are interconnected, the faster the rate of adoption. Fourth, is the influence of the extent of the change agents' (e.g., advertisers) promotional efforts. The greatest payoff for the advertisers' efforts occurs when opinion leaders are adopting.

The Rogers model has been criticized on the grounds that adopter categories are not consistent, meaning that individuals tend to be innovators in certain areas and laggards in others (Kotler 1991, p. 343). This means the after-the-fact classification of adopters into categories becomes tautological and lacks predictability.

Figure 6.2 **Distribution of Product Adopter Categories**

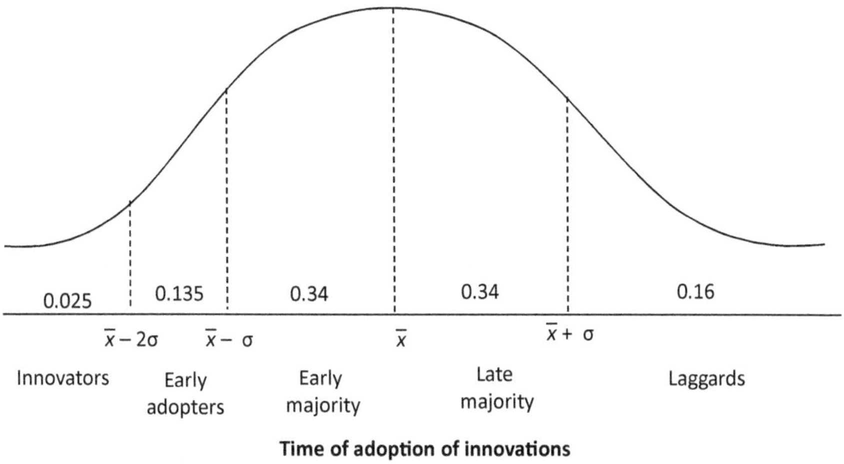

Source: Adapted from Rogers (2003).

Figure 6.3 **Cumulative Diffusion Curve**

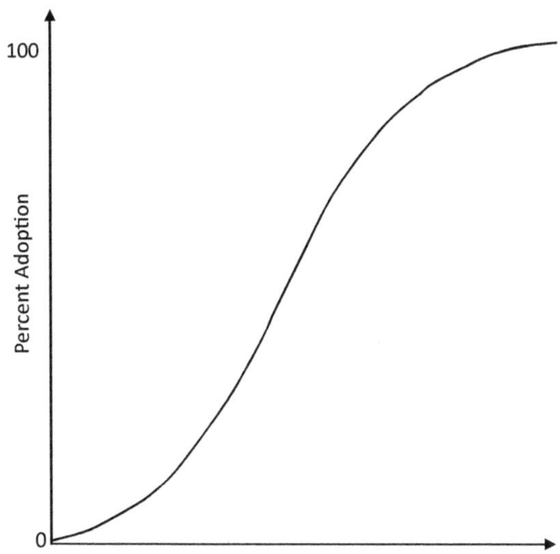

THE BASS MODEL

While the Bass (1969) model is similar to the Rogers model in that the cumulative adoption curve reflects both interpersonal (word of mouth) and mass media communication, the former has become the standard model of diffusion because of its greater predictability. The Bass model assumes that there are two groups that comprise potential adopters of an innovation. Potential adopters are segmented according to the type of communication by which they are influenced. One group is influenced by mass media (external influence) and labeled "innovators"; the other group is influenced by word of mouth (internal influence) labeled "imitators."

Note that Rogers's model in the fifth edition (2003) of his book also considers communication channels—mass media versus interpersonal—as an important determinant of the rate of adoption. What is different about the Bass model is that it unbundles into two parts—those who adopt due to internal influence and those who do so due to external influence.

The basic Bass model is a first-purchase diffusion model. That is, there are no repeat buyers, or what Bass calls "replacement sales," and each buyer purchases one unit. As a result, the number of adopters is equal to the number of unit sales throughout the diffusion process. In the Bass model, initial purchases are made by both "innovators" and "imitators." The distinction between the two is that the former are not influenced in the timing of their initial purchases by the number of people who have already bought the product, while the latter are influenced by the number of previous buyers. Imitators learn from previous buyers prior to purchase.

As shown by Mahajan, Muller, and Bass (1990), the basic Bass model can be expressed as:

$$n\,(t) = p[m{-}N(t)] + q\,\frac{N(t)}{m}[m{-}N(t)]$$

Where at any time t,
$n(t)$ = noncumulative number of adopters at time t
$N(t)$ = cumulative number of adopters at time t
m = total market potential for a new product
p = coefficient of innovation or external influence
q = coefficient of imitation or internal influence

Figure 6.4 illustrates graphically the structure of the Bass model. Mahajan, Muller, and Srivastava (1990) cast Rogers's five adopter categories in terms of the Bass model and provide a comparison of the two models shown in Table 6.2. Note that because the Bass classification is innovation specific, the percentage of adopters in each category varies. Moreover, the innovator category in the Bass model appears at the start of the diffusion process.

Internal and External Influences

In a meta-analysis of 213 applications of diffusion models from fifteen articles, Sultan, Farley, and Lehmann (1990) found that the diffusion process is much more affected by imitation (adoptions due to internal influence) than by innate innovativeness (adoptions due to external influence) of the consumers. In the context of the Bass model, this means the average of the coefficients of imitation (q) is several times larger than the average of the coefficients of innovation (p). That is, the influence of word-of-mouth communication (internal influence) on diffusion is much more important than that of mass media communication (external influence). This finding is consistent with the Bass model, which assumes that mass media is

Figure 6.4 **The Bass New Product Diffusion Model**

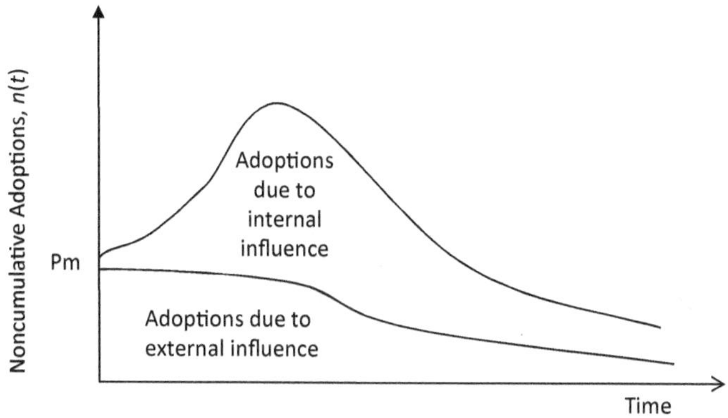

Source: Mahajan, Muller, and Bass (1990, Figure 1.1 A, p. 5). Reprinted with permission from the *Journal of Marketing Research*, published by the American Marketing Association.

Table 6.2

Comparison of Bass Adopter Distribution with Rogers Normal Distribution

		Bass Adopter Distributions	
Adopter Category	Normal Adopter Distribution (%)	Range (%)	Approximate Stable Value After *q/p* = 40 (%)
Innovators	2.5	0.2 to 2.8	1.0
Early adopters	13.5	13.3 to 20.2	18.0
Early majority	34	29.1 to 32.1	29.5
Late majority	34	29.1 to 32.1	29.5
Laggards	16	21.4 to 23.5	22.0

Source: Mahajan, Muller, and Srivastava (1990, p. 44). Reprinted with permission from the *Journal of Marketing Research*, published by the American Marketing Association.

important in the early part of diffusion process, but interpersonal communications are more important as time passes.

Mahajan, Muller, and Wind (2000) document numerous applications of diffusion models involving strategic decisions. These decisions include pre-launch decisions (e.g., forecasting of new product sales and product sampling); post-launch decisions (e.g., timing of successive generations of a product); estimation of lost sales and market expansion due to patent infringement, the impact of pricing on diffusion, and many more.

DIFFUSION FROM AN ECONOMIST'S PERSPECTIVE

Hall (2004) provides a list of factors classified into five main groups that influence the diffusion of innovators from an economic perspective. These factors are (1) benefits received from the new technology, (2) network effects, (3) costs of adopting the new technology, (4) industry or social environment and uncertainty, and (5) information problems comprising these five groups. Hall identifies these groups with the Rogers's classification system—the first and second groups with relative advantage and complexity, the third with compatibility, and the fourth with trialability and observability. Below is a description of Hall's five groups of factors.

1. **Benefits of new technology.** These benefits are rather small in the beginning, but increase over time because of the feedback between diffusion and innovation. That is, technology is improved as users learn and provide feedback to the suppliers (e.g., applications software). The rapid rise of open source software has been dependent on just such user feedback. Most models of diffusion, including the ones discussed here, do not consider the feedback effect.
2. **Network effects.** As discussed in chapter 3, network effects arise when the value of some new product or service to each consumer increases as more and more consumers adopt it. Such products or services (e.g., Internet or computer operating systems) are said to produce positive network externalities. As noted in chapter 3, network effects—also referred to as demand-side economies of scale or bandwagon effects—are present when consumers of a product benefit from greater availability of complementary products (e.g., software for users of personal computers). The presence of network effects has several implications for product as well as pricing and promotional strategies. Other things being equal, the presence of network effects provides innovating firms with strong incentives to formulate and implement strategies that result in rapid diffusion in order to benefit from first-mover advantages. As discussed in the section on design issues, the objective is to quickly build a large installed base through a variety of product, pricing, and promotion strategies. Standards play a large role here because they help create network externalities facilitating the consumer learning process, by increasing the availability of complementary products and the like. Moreover, with the presence of network externalities in a product, consumers demand a common standard (this need not always be a single standard). Overall, network effects, other things being equal, tend to accelerate diffusion of new products.
3. **Relative costs of adopting the new technology.** The greater the relative costs of adopting the new technology, the slower the rate of adoption, other things being equal. Such costs include the direct costs of acquiring the new technology and the complementary investment in the form of necessary capital equipment and training as well as the indirect costs associated with learning required to use the new technology. As a result, it is not uncommon for innovator firms to provide free training and incentives to make the technology attractive to potential adopters.
4. **Information and uncertainty.** The greater the knowledge required to learn about a new technology, and the greater the uncertainty associated with it, the slower the rate of adoption. New and complex technologies and products impose high costs of training and learning on potential adopters, and therefore increase their relative costs. Moreover, potential adopters can reduce

the uncertainty about the benefit of a new technology by simply delaying the adoption. Both of these factors result in slower rates of diffusion. In addition to promotional strategies, which emphasize education of consumers, user involvement from the early stages of development of new technologies could mitigate the problem of information uncertainty. The uncertainty of the benefits of the new technology could be reduced by sharing the risks of new technology through a variety of guarantees and warrantees.

5. **Industry environment.** A market structure of few sellers can result in higher prices, slowing the rate of diffusion while at the same time producing a common standard quickly, increasing the rate of adoption. Similarly, regulation can increase or slow the rate of adoption.

INNOVATION BY USERS AND THE ROLE OF LEAD USERS IN DIFFUSION

Innovation by users has been the subject of a relatively small, but growing body of literature in the past three decades with important implications for diffusion. Research in this area owes much to the contributions of Eric von Hippel (2005) and his collaboration with others—Urban and von Hippel (1988) and Franke and von Hippel (2003a and 2003b), for example. Von Hippel argues that user innovation results from heterogeneity of user needs—requiring a high degree of customization—not served by the market. This is evidenced in a number of studies that reported high variance within segments.

Von Hippel (2005) documents survey results in eight product areas where the percentage of users reporting developing or modifying products for their own use ranged from 19 percent to 36 percent for five industrial products and 10 percent to 38 percent for three consumer products. Moreover, numerous studies found that some of the most important and novel products and processes have even been developed by users—both user firms and individuals. They include most of the innovations in oil refining (Enos 1962); the most widely licensed chemical production processes (Freeman 1968); about 80 percent of the most important scientific instruments (von Hippel 1988); and the most commercially important equipment innovations in sporting fields (Shah 2000). More recently, evidence from 1981–2000 U.S. patent data suggests that, in the information and communications technology (ICT) sector, the locus of problem solving is shifting away from firms specializing in that sector to direct users of a product or a service, many of whom are non-ICT firms (Rao, Vemuri, and Galvin 2004).

In the traditional model of innovation, information about what the user/customer wants or needs often lies with the customer, and information about how to satisfy the customer's needs lies with the manufacturer. Users are forums or individual consumers that expect benefit from the use of a product or service. In contrast, manufacturers expect benefits from the sale of a product or service (von Hippel 2005). In the traditional model, serious interface with the customer does not occur until the prototype is built. In contrast, in the user-based innovation model, interface starts at the design stage. Sharing of otherwise proprietary information through packaged tool kits that allow customers to be intimately involved at the design and prototype building stages is the key to implementing user-based innovation.

The Role of Lead Users

Von Hippel (2005) notes two distinguishing characteristics that separate lead users from the rest of the user population:

1. They are ahead of an important market trend and they are in need of solutions before many other users;
2. They expect to derive relatively high benefits from obtaining a solution to their needs, so they are better equipped to innovate.

Morrison, Roberts, and Midgley (2004) found significant correlation between the two defining characteristics of lead users and the likelihood that they will innovate. Moreover, as shown in Figure 6.5, Franke and von Hippel (2003a and 2003b) found that the higher the intensity of lead user characteristics displayed by an innovator, the greater the commercial attractiveness of the innovation the lead user developed. Also, Lüthje (2004) found that the speed of adoption is faster among innovating consumers compared with noninnovating consumers.

Figure 6.5 **Relationship Between "Lead-user-ness" and Attractiveness of Innovation**

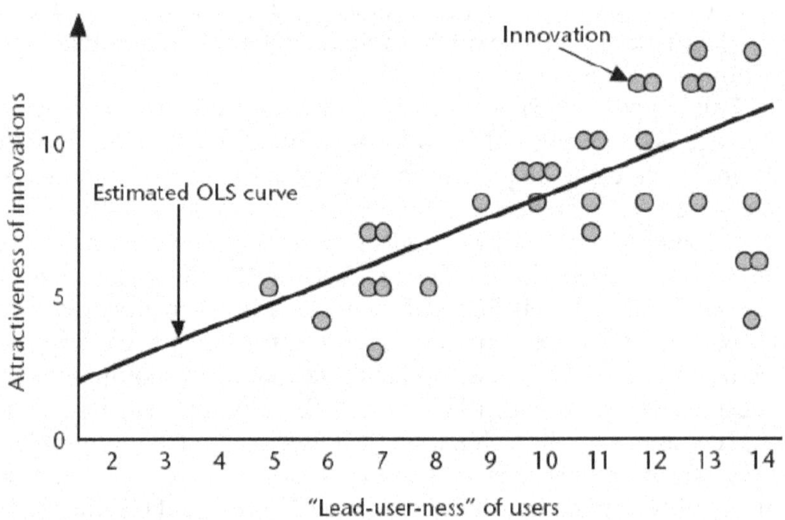

Source: Von Hippel (2005). © 2005 Massachusetts Institute of Technology, by permission of MIT Press.

Google, as a major user of open source software (OSS) in its internal systems as well as in its online services, fits this definition of lead user well. Lilien et al. (2002) report that concepts for new products developed by lead user project teams had far more commercial promise than those developed by non-lead-user teams. Note that lead users do not necessarily produce the product themselves. They are clearly the originators of the concept and basic design of the product and often collaborate with the manufacturers. Unlike the diffusion process in the Rogers and Bass models where "innovators" are merely the earliest adopters, in the user-innovation model they are innovators as well as lead users. As such, they not only help speed up the diffusion process but also gain the ability to set standards for the product in question.

NETWORK EFFECTS AND DIFFUSION IN PHARMACEUTICAL MARKETS

Network effects exist in several different markets. The high-tech pharmaceutical market provides a clarifying example. Berndt, Pindyck, and Azoulay (1999) examined the role of network effects in pharmaceutical markets with focus on antiulcer drugs. Network effects in the pharmaceutical market arise when the demand for a branded drug by patients and physicians depends on the number of other patients that have taken or are taking the drug. That is, the network effects are mainly informational in that the greater the use of a drug by others, the greater the influence on one's perception about its efficacy, safety, and the like. This is not unlike the bandwagon effect, which results in a particular drug—not necessarily the best—dominating the market, thus affecting its valuation and rate of diffusion.

The authors found network effects at the level of brand (e.g., Zantac versus Tagamet) as well as at the level of entire therapeutic class (e.g., a class of antiulcer drugs) with important strategic implications. Pioneering firms benefit by establishing a large installed base before other firms enter. As more people use the product, word-of-mouth communication increases, accelerating the rate of diffusion.

PLATFORM STRATEGY

At the simplest level, a platform strategy involves using a common base of components or technology—"a platform"—upon which to build multiple models. When a product is largely proprietary and under control of one company, the strategy is referred to as "in-house" product platform strategy (Gawer and Cusumano 2008). The idea is to implement a differentiated marketing strategy (i.e., targeting multiple segments) while at the same time minimizing the loss of scale economies in production. Put differently, a product platform strategy enables firms to offer greater variety without incurring an undue cost penalty. A key element in such a strategy is the implementation of modular product architecture where loosely coupled component interfaces are highly standardized and specified such that a change in the design of one component—in contrast to the traditional design process—does not lead to changes in other component designs. Sanchez (1999) suggests that modular architecture makes it possible to create greater variety and to introduce new products to market more rapidly, all at a lower cost. He offers PC as an example where the substitution of variations in any one component—disk drives, memory cards, monitors, keyboards, and so on—within the PC architecture occurs with relative ease without changes in other components.

In contrast to product platform strategy, a firm that seeks to develop and "own" an industry platform—upon which a wide range of interdependent businesses operate—has less control of its own intellectual property. This is because success in platform leadership depends on innovation in complementary products and vice versa. Such, for example, is the case with Microsoft's Windows operating system, which needs a continuous stream of innovation by firms specializing in applications software and peripheral devices (i.e., complementors) for its success as an industry platform leader. This requires Microsoft, which is otherwise known as a firm that follows a closed proprietary model of innovation, to be collaborative and open its technology to complementors.

NETWORK EFFECTS

Gawer and Cusumano (2008) and Cusumano (2010) suggest that a critical feature of an industry platform is the creation of "network effects." These effects are either "direct" effects from technical compatibility of an interface standard or "indirect" effects from a large number of players (including application developers and content producers) adopting a particular platform. An industry platform with network effects leads to more users to adopt the platform, which in turn leads to more users and complementors. This is akin to bandwagon effects, discussed in chapter 1.

Gawer and Cusumano (2008) provide many examples of companies following platform strategies in many different industries. They include Qualcomm, which established its wireless communications technology as a platform for the cell phone industry, Google's Internet search technology as a platform for advertising, and many more. Isckia and Lescop (2009) provide a fascinating view of Amazon's evolution starting from online retailer of books to creating an Amazon platform, which provides e-services (see Exhibit 6.1).

Exhibit 6.1 **The Amazon Case**

In 1994, Amazon began with the online Amazon Bookshop offering great value to its customers in terms of high-quality service, convenience, and competitive pricing through innovative use of technology. Within two years, Amazon launched its Amazon Associates as a means of increasing its sales.

The relationship with the Associate is one in which Amazon pays the Associate for links from the Associate's site to Amazon's site that result in actual sale. During the next stage, from 1997 through 2004, Amazon expanded the e-book selling platform into an e-commerce platform, allowing businesses to offer their products ranging from music, to computers, to shoes on its Web site. Then, in 2005, Amazon began expanding the scope of its Web services into delivering storage capacity (Amazon S3), computing power (Amazon EC2), E-Commerce service (ECS), and the like.

Thus, within a period of just over ten years, Amazon created and sustained one of the most successful online retail industry platforms in the world and leveraged its technology to move into providing Web services to third-party retailers. As a leader of this industry platform, Amazon met many of the prerequisites suggested by Gawer and Cusumano (2008). These include solving fundamental technical and market problems associated with developing a successful industry platform, making the platform easy to connect to and build upon, encouraging complementary innovations by opening up its own proprietary system, and providing economic incentives to partners.

PLATFORM STRATEGY AND THE CASE FOR OPEN SERVICES INNOVATION

Chesbrough (2011) argues that product-driven companies are finding it harder to differentiate, much less sustain their strategy over time because of rapid commoditization of products brought about by wide availability of best practices (e.g., Six Sigma and customer relationship management) to everyone in software form. Chesbrough's solution is to transform physical products into platforms with a variety of services created by open innovation. While product-driven firms will continue product innovation, competition would shift toward the platform of products and services—many of these produced by unaffiliated firms—available to customers through their products. Chesbrough (2011) contrasts the limits of product-focused innovation with the benefits of platform-driven open services innovation by examining the case of Motorola's once highly successful RAZR cell phone. This revolutionary phone fell by the wayside within three years because of pressure from the likes of Apple and Google, who are building platforms of applications and services surrounding their cell phones in collaboration with thousands of companies to create and sustain competitive advantage in the product as well as services market.

Note that open services innovation is part of what some scholars believe is a larger trend toward the open innovation (OI) model, which goes against the traditional approach to maximizing profits from innovation through strong IPRs (see Christensen 2006, and West, Vanhaverbeke, and Chesbrough 2006). The essence of the OI model, in contrast to the traditional closed and vertically integrated model of innovation, is that it enables firms to receive external ideas and produce competitively superior innovation and commercialization by opening their own innovation to outsiders. As such, it challenges the core competency perspective of corporate innovation and emphasizes research and productivity via collaboration and networks (Ghauri and Rao 2008). As discussed in chapter 1, Pisano's (2006) story about Merck's collaboration with Washington University to create a database for the human gene sequence—the Merck Gene Index—and place it in the public domain illustrates this point. This is because the OI model is designed, among other things, to increase the quantity and quality of the innovation pie, from which the innovating firm and its collaborators derive greater benefits than would be the case otherwise. Moreover, a firm like Merck with strong proprietary interest in the outcomes of human gene sequence basic research may find it strategically attractive to prevent privatization of genes that could block its future research.

BRANDING

Successful branding of a product or service ranks among the highest sources of sustainable competitive advantage. This is because, unlike the design of a product or its production process, a successful brand name cannot be imitated, although its competitive position can be challenged by a generic product (e.g., generic drugs) or by another powerful brand (e.g., Zantac versus Zocor). Branding is a critical component of product strategy. It is a source of differentiation in the sense that it enhances the relevance and value of the product to the target market. Table 6.3 shows the ratio of brand value to assets for companies that comprise the top twenty brands in 2010, where brand value comes from data published by Interbrand. Note that an overwhelming number of these company brands are high-tech and that the brand value is at least 50 percent of assets for many of the companies. While marketing literature on branding is long and deep, our task in this section is to address two aspects in the context of product strategy: first, we provide a brief discussion on the concept of brand equity

Table 6.3

Ratio of Brand Value to Assets of Selected Companies, 2010 (dollars in millions)

Brand	Brand Value	Assets	Ratio of Brand Value to Assets
Coca-Cola	$70,452	$72,921	0.966
IBM	64,727	113,452	0.571
Microsoft	60,895	86,113	0.707
Google	43,557	57,851	0.753
General Electric	42,808	751,216	0.057
McDonalds	33,578	31,975	1.050
Intel	32,578	6,316	0.507
Nokia	29,495	51,846	0.569
Disney	28,731	69,206	0.415
HP	26,867	124,503	0.216
Toyota	26,192	327,408	0.079
Cisco	23,219	81,130	0.286
Louis Vuitton	21,860	49,250	0.444
Apple	21,143	75,183	0.281
Samsung	19,491	119,809	0.163
Honda	18,506	125,455	0.148

Sources: Mergent Online, Hoover International, and Interbrand (2011).

developed by David Aaker in his many publications (1991 and 1996, for example), and including Aaker and Jacobson (2001) and second, we consider special issues in branding of innovations.

THE CONCEPT OF BRAND EQUITY

Aaker (1991) defines brand equity as "a set of brand assets and liabilities linked to a brand, its name and symbol, that add or subtract from the value provided by a product or service to a firm and/or to that firm's customers" (p. 15). Later, Keller (2008) defined customer-based brand equity as "the differential effect that brand knowledge has on customer responses to the marketing of that brand" (p. 48). Note that the underlying assumption in both definitions has to do with the differential effect of brand as providing value to the customer and to the firm.

Aaker (1991) identified five categories of assets/liabilities that underlie brand equity and are linked to the name and/or symbol of the brand: (1) brand loyalty, (2) name awareness, (3) perceived quality, (4) brand associations (e.g., the favorable link of Ronald McDonald to McDonalds's brand), and (5) other proprietary brand assets such as patents and trademarks, which provide some legal protection against imitation. Overall, these assets work to provide value to the customer in terms of user satisfaction, confidence in purchase decision, and ease of information processing. Greater value to the customer in turn leads to a greater value to the firm in terms of efficient and effective marketing programs, premium prices, brand extensions, trade leverage, and the like. In short, a well-managed brand contributes to sustainable competitive advantage.

BRANDING INNOVATIONS

Aaker (2007) notes that, despite volumes of books and articles written on the subject of innovation, there has been little discussion on how innovation should be branded. He suggests three ways that branded innovations can help a firm:

1. They help exploit or appropriate the economic benefits from a new or improved product by making it more differentiated. Without branding the product could become a commodity. Aaker suggests that branded innovation allows the firm to "own" the product and helps withstand competitive pressure; adds credibility and legitimacy by signaling its worth; enhances visibility and awareness; and aids in communicating complex features of a product.
2. They help create a subcategory (e.g., iPod as subcategory of portable MP3 players).
3. They help create and sustain the reputation of the firm for innovativeness and enhance its overall corporate brand.

However, not all innovations present an opportunity to successfully create a new brand. Aaker suggests three simple but important criteria to help decide whether the innovation is worth branding: the innovation must be significant and superior to existing offerings; customers must care enough to change their loyalty from other brands; and it must merit complementary investments in promotion, distribution, and marketing assets. We would add that branding becomes necessary in the context of easy replicability and/or weak intellectual property rights. That is, branding becomes an important means by which high-tech firms protect themselves against a weak appropriability regime.

Generic Branding Decisions

Keller and Lehmann (2006), in an important review of literature on brands, identified several key generic branding decisions and tasks that apply to all products. Of these, decisions concerning brand positioning, evaluation of brand performance, brands as a source of firm growth, and strategic management of brands are particularly relevant to high-tech markets.

Developing Brand Positioning. This involves differentiating the brand and establishing competitive superiority. Branding decisions that are particularly relevant to high-tech markets include tangible product and intangible attributes as well as corporate image and reputation. The latter is particularly important in high-tech markets. For example, Keller and Aaker (1998) have shown corporate marketing activities that help establish the reputation and credibility of a firm as an innovator, and those activities that are favorable for brand extensions.

Evaluation of Brand Performance. This has to do with measuring brand equity from three different perspectives—the consumer perspective, the company perspective, and the financial market perspective. From the consumer perspective, brand equity involves the value customers place on the "nonobjective" part of the product offering by the firm. Keller and Lehman (2006) suggest that the customer brand equity can be captured essentially by many of the same aspects (e.g., awareness, associations, loyalty, and the like) identified by Aaker (1991) and noted earlier. However, with respect to high-tech products and services it is difficult to separate the nonobjective part from the objective part of the product offering. Difficulty arises because the novelty of technology combined with the highly attractive product attributes technology generates (e.g., iPad) become dominant and synonymous with the brand.

Brand equity from a company perspective is the incremental discounted net present value (i.e., net of investments associated with building a brand) that the firm derives from the brand. This is manifested in premium pricing, increased revenue elasticity of advertising, easier access to distribution, and increased barrier to competitive imitation—all of which result in greater appropriability of innovation. In terms of the financial-market perspective, the market value of the firm should reflect brand equity, which is the net present value of future cash flows. Note that under conditions of efficient financial markets with no errors of measurement, brand equity measured in terms of expected net present value of future cash flows would be equal to the financial valuation of the brand reflected in the stock price. However, measurement of brand equity at the firm level and the financial-market level in the context of high-tech products and services is compounded by the fact that a high-tech firm typically invests in considerable sums of R&D, holds patents and copyrights, and employs a significant amount of human capital assets. All these items fall in the category of intangible assets, but none show up as such on the balance sheets. Quantitatively separating brand equity from the effect of these various other intangible assets is extremely challenging. Still, a wide range of methodologies and models for measuring brand equity at the firm level, at the financial-market level, and at the customer level exist. The reader is referred to Christodoulides and Chernatony (2010) for an excellent review of literature on conceptualization and measurement of customer-based brand equity. The reader is also referred to Keller (2003), Keller and Lehman (2006), and Simon and Sullivan (1993) for a sample of various methodologies.

Brands as a Source of Firm Growth. The standard growth strategies, as discussed in chapter 2, emphasize four ways for firms to grow—(1) greater penetration of the existing market; (2) developing new products for the existing market; (3) entering into new markets with an existing product; (4) and diversification into new markets with new products. In this framework, new product development and new market development create opportunities for new brands, which in turn contribute to additional growth through line and category extensions. Therefore, branding not only helps insulate high-tech firms from imitation, but also becomes an important source of growth. Put differently, branding helps high-tech firms achieve greater returns on their investments in innovation than would be the case otherwise.

Strategic Management of Brands. Technology markets differ from other markets in many respects, with important implications for brand management. Sawhney (2005) suggests that, in comparison to markets for consumer packaged goods, technology markets change rapidly, leading to branding the company rather than the product. In addition, technology markets have (1) shorter product life cycles, requiring brand positioning across successive product generations; (2) greater product complexity with many features emphasizing the product rather than the brand; (3) more complex and modular product architecture with several components and many suppliers, with emphasis on brand architecture in terms of extensions, sub-brands, co-branding, and ingredient branding; (4) more complex customer decision-making processes with multiple sources of information, meaning that management of a brand is a total customer experience; and (5) vertical as well as horizontal channel and partner relationships (e.g., complementors, value added resellers [VARs] and independent software vendors [ISVs], leading to conflicts of interest with management. Sawhney argues that, in general, technology firms tend to focus on branding the company and tend to rely more on the "branded house" or the "sub-branding" approach compared to the "house of brands" approach favored by the consumer packaged goods firms like Procter & Gamble.

Brand Versus Reputation

As noted earlier, Etterson and Knowles (2008) suggest that brand and reputation, although related, are not one and the same, and confusing the two can lead to costly mistakes. The authors maintain that brand is a "customer-centric" concept that focuses on the promise of a product or service to the customer, whereas reputation is a broader, "company-centric" concept with focus on the credibility an organization commands among a wide range of stakeholders, including employees, investors, communities, as well as customers. Thus, while the reputation of a firm depends, in part, on being a good citizen, the strength of a brand depends almost entirely on the benefits it provides to the customer.

However, what is more interesting is the interrelationship between brand and reputation. One would expect the relationship to be especially strong in mono-brand companies, that is, where the name of the company and brand are one and the same (e.g., IBM and Google). In such cases, good (or bad) reputation can produce positive (negative) spillovers to the brand, and vice versa. In the context of high-tech markets, a firm with a strong reputation for producing innovative and reliable products would find it easier to develop new brands and defend existing brands than would be the case otherwise. Similarly, a significant brand that fails can damage the firm's reputation for innovation and its ability to introduce new products and develop new brands. Roberts (2011), based on his study of the online tractor market where quality is imperfectly observed or delivery is uncertain, finds that warrantees like a "guaranteed or your money back" promise from the producer do not substitute for reputation in either influencing the price or the probability of sale.

Ingredient Branding

Creating a brand name for complementary ingredients—materials, components, parts, and the like—contained within end products is called ingredient branding. The strategy of ingredient branding involves taking a B2B (business-to-business) product—a product that is an input into making another product—to the final consumer, which then becomes a B2C (business-to-consumer) product. Some well-known ingredient brands include Makrolon, a plastic manufactured by Bayer Material Science AG that is widely used in everyday products ranging from water bottles to CDs to storage boxes for food; Techron, a fuel additive made by Chevron; and the Intel microprocessor used in personal computers. Ingredient branding is particularly relevant for high-tech markets because it is often a technological innovation, in the form of new and improved complementary input, which gives rise to such a strategy. It is not surprising that virtually all of the more than 150 ingredient brands listed by Kotler and Pfoertsch (2010) can be traced to significant high-tech inputs that go into final products. This is because technology-intensive inputs that are central to the success of the final product make good candidates for ingredient branding. However, there are several conditions for success in ingredient branding. They include:

1. Consumers must perceive the ingredient brand as central to the performance of the end product, like the Intel microprocessor for the personal computer.
2. The branded ingredient is highly differentiated and the consumer is convinced of its sustainable value (e.g., Bose with its sound system).
3. The company making the final product supports and seeks the ingredient brand in order to differentiate its product offering. This is especially the case with weak or relatively unknown downstream brands. Chinese heavy construction equipment companies using Cummings diesel engines is a case in point.

The benefit to the company producing the end product depends on the extent to which the ingredient brand brings additional value to its brand. If it brings value, volume and prices will be greater than they would be otherwise, especially if the end product's brand is not stronger. In this case, the supplier of the ingredient brand benefits from greater bargaining power in terms of favorable price, less competition, and so forth. On the negative side, the supplier becomes directly accountable for the quality and has to incur the cost of promoting the brand. Similarly, the risk to the firm producing the end product involves compromising its own brand-building effort.

The benefits to the final consumer from ingredient branding include greater confidence in the purchase—especially when the end product's brand is weak—and ease of information processing as is the case with any successful downstream brand. On the other hand, if the end product itself carries a strong brand name (e.g., Sony), there is no need to add the ingredient brand (e.g., Dolby sound technology) to its package. Finally, the degree of technological complementarity that exists between the end product and the ingredient in question matters.

PRODUCT STRATEGY AND THE MARKET FOR TECHNOLOGIES

During the past two decades, there has been rapid growth in what has come to be called the market for technologies, as opposed to physical products in which technologies are embodied (Baumol 2002; Arora, Fosfuri, and Gambardella 2001). This is evidenced by the rapid growth—about 9 percent per year—in U.S. receipts and payments of royalties and license fees over the 1999–2009 period (see Table 1.1 in chapter 1). Note that the receipts and payments associated with industrial processes and general software—where only licensing fees are involved—have been growing at a phenomenal rate of 25 percent per year.

As discussed in chapter 3, the market for technologies involves buying and selling disembodied technologies at a price. The transaction could involve the outright sale of the technology, licensing, cross-licensing, and the like. Licensing appears to be the chief means by which technologies are marketed. Note that licensing involves balancing increased revenues from licensing against the lower profits because of potential imitation by the licensee. License fees and the associated terms of the contract constitute the price at which the technology is marketed. Although such transactions occur largely in the B2B market (e.g., chemical processes), a growing number of applications software "products" can be just as easily described as "technologies" in the B2C market. Licensing works when IPRs are strong and technology is sufficiently codified.

LICENSING TECHNOLOGIES

Arora, Fosfuri, and Gambarella (2001) note several factors that explain the growing licensing market for technologies. They include increased demand for technologies from globalization, which allows firms to exploit technology on a large scale in a relatively short time; improved opportunities for licensing because of improved organization and functioning within the market; and more efficient management of the high cost of R&D and other intellectual capital. Two other factors should be added to this list—a growing tendency to view the "R&D lab" as a profit center, and the use of licensing as a strategic bargaining chip in the formation of alliances and other competitive strategies.

A significant factor in product strategy is timing. In terms of product-related factors influencing the incidence and timing of technologies, Kotabe, Sahay and Aulakh (1996) conceptualized the following propositions:

- The installed base of components and end products is positively related to network externalities, which in turn leads to greater technology licensing.
- A large number of interfacing components and compatibility requirements lead to greater network externalities and technology licensing.
- When technical standards are less established, the effect of compatibility requirements on the incidence and timing of technology licensing is stronger.
- In terms of resource-based factors, higher incidence of licensing is associated with a weak appropriability regime, lower degree of complementary assets, and higher development costs. When there is a greater emphasis on speed to market, then the incidence of technology licensing is higher. The authors also examined the influence of technology licensing on product strategy.
- Higher technology licensing leads to greater standardization of end product and greater modular upgradability at the component level, which in turn leads to higher network externality to the customer and lower compatibility requirements.

- The higher the level of licensing, the greater the product overlaps.
- A multimarket strategy leads to higher-level technology licensing than a single-market strategy. When a firm such as IBM offers products and/or components in multiple end-product markets, it has a multimarket strategy in contrast to a single-market strategy where a firm's offerings of products and/or components are aimed at a single end-market.

APPROPRIABILITY

In what respects is the market for technologies different from the market for products? And what do the differences mean for "product strategy" in the technology market? In general, technology markets suffer from the appropriability problem to a greater extent than the market for high-tech products. This is because technologies, especially those that are codified, such as software programs, are easily copied and shared with or without copyrights and patents. As a result, the potential buyer cannot be sure she/he has exclusive use of the technology and the company cannot be assured that its product will be protected from imitators.

Moreover, there is a strong element of moral hazard in transactions involving technologies. That is, often the seller has to disclose important information about the technology to the potential buyer before the purchase, and once this is disclosed, the buyer, in the extreme case, has no obligation to buy the technology even with a nondisclosure agreement. A good example of this is the software source code underlying a proprietary operating system like Windows or Macintosh. Alternatively, the buyer may negotiate for a lower price. If the potential buyers needs information in order to make a decision and the seller refuses to disclose that information, then the transaction will not occur. Opportunistic behavior resulting from information asymmetry between the potential buyer and the potential seller—the latter has all the information about the technology in question—is at the center of this problem. As discussed in chapter 1, these problems, combined with a high level of uncertainty associated with market transactions involving new technologies and the difficulty of writing contracts, present great challenges to marketers.

It should be noted here that appropriability of technology varies depending on whether it is product or process technology. Process technology, which has to do with the manufacturing or production process, is particularly difficult to imitate and therefore appropriable to a greater extent, despite the fact that many such processes are not patentable. Pisano (1997) suggests that firms in such industries as advanced materials, specialty chemicals, and semiconductors leverage in-process technology development to achieve accelerated time to market through rapid production ramp-up and enhanced acceptance of new products, leading to greater appropriability of product innovation.

APPLICABILITY OF GENERIC MARKETING STRATEGIES

In light of the above description of the market for technologies, the key elements of normal marketing/product strategy—market definition, segmentation, targeting, positioning, and managing brand equity, for example—while still applicable, need to be adapted. Two important articles—Capon and Glazer (1987) and Glazer (1991)—provide some insights on this issue. They include:

- The market boundaries for technologies are difficult to define initially and they shift rapidly, making segmentation and targeting difficult.
- Even when segmentation is carried out, the rapidly changing definition of the market makes it less stable, throwing initial targeting and positioning strategies into disarray.
- Rapid technological change combined with equally rapid changes in market structure make it difficult to keep a firm's technology proprietary. As a result, the market life cycle of the particular technology firm becomes shorter.
- Unlike distribution of physical products, which involves elaborate management of channels, including logistics, distribution of technology is akin to communication of valuable information, which is difficult to protect from leakage. As a result, distribution of technology is direct and typically involves two parties, the buyer and the seller. Still, while the direct cost to the seller of reproducing the technology is relatively small, the indirect cost of *transferring*—time expended by employees in the process, for example—technology to the buyer can be considerable. Terms of sale and licensing, such as geographic and field-of-use segment restrictions, dominate distribution problems.
- With respect to pricing, major decisions involve the valuation of technology, as discussed in chapter 3. Given that the demand for technology is derived from the demand for downstream buyers' products and services, a key question for identifying the value of technology centers around the economic benefit the buyer derives from the application of technology. Other generic questions concerning the valuation of technology concern the strength of legal protection in terms of patents and copyrights, the threat of substitute technologies, the threat of backward integration by the buyer, and the availability of complementary technologies.

THE ROLE OF BRANDING

It might seem, at first blush, that branding and reputation play little or no role in the marketing of technologies, which after all are "invisible products," and that rational buyers make the purchase decision based on objective evaluation of available information. However, we argue that, in light of information asymmetry and other impediments to normal market transactions noted above, branding and reputation have a large role to play even in the market for technologies. The concept of ingredient branding, discussed earlier, offers the most promising avenue for branding technologies. The success of the Android operating system is a case in point. It is a case of developing and marketing a branded *technology* as opposed to a physical product. Consumers talk about Android phones, not about Motorola phones with the Android operating system (see Bishop 2011). Reputation of the firm for innovation, quality, and *trust* becomes especially important in marketing technologies because, as noted earlier, reputation, unlike a brand name, is a statement about the entire firm that goes beyond its products. Bidault and Fischer (2007) found that given the very high transaction costs combined with incentives for opportunistic behavior associated with marketing of technologies, firms wishing to acquire new technologies turn first to trusted business partners, and that successful technology transactions typically take place between parties who had a prior relationship (see Noteboom 1996). Noteboom notes that the relational view of transactions has long been advocated by marketing scholars like Johanson and Mattson (1987) and Morgan and Hunt (1994), who argue that trust is central to relationship marketing.

The Role of Quality Certification

As discussed previously, building brand equity is one of the chief means by which firms in technology markets fight against imitation and achieve sustainable competitive advantage. Perceived quality is one of the major assets that underlies brand equity. However, quality is not always observable. This is especially the case with process-oriented attributes such as quality of engineering practices in the software development projects. Such attributes are often referred to as "Potemkin" attributes, which are characterized by the fact that neither the buyer nor the external institutions are able to conduct laboratory analysis at the end-product level. This leads to a high degree of information asymmetry between the supplier and the customer—a source of potential market failure—which is not easily overcome by the traditional quality signals communicated by branding, guarantees, warrantees, and complementary advertising. Certification is a means by which information asymmetry is reduced and quality assured, provided the certifying organization succeeds in establishing a reputation for high standards for itself "Certification" becomes a public good, which mitigates the potential for market failure arising from information asymmetry. Note that certification may not cover all dimensions of quality. Firms that have certification compete by differentiating themselves in other ways, including branding.

The standards developed by the International Standards Organization (ISO)—the ISO 9000 production standards and the ISO 9001 for software—have become widely adopted international quality standards. ISO certification requires the firm to have well-defined, well-documented, and verifiable processes that it follows. Once the certificate is awarded, companies use it as a marketing tool to promote their products. ISO-type certification is especially important for newer and smaller firms from developing countries trying to break into the developed world's markets.

The Case of the Indian Software Industry. The results of a study by Arora and Asundi (1999), using data on 95 Indian software firms and their U.S. clients, indicate that ISO certification increases the quality of a firm's output, enhances revenue growth, and enables firms to receive higher unit price than they would otherwise. The authors also found that most firms see ISO certification as an important marketing tool. However, it is not a substitute for branding the innovation. Firms that go through the required process and obtain the certification use it as a means by which to differentiate themselves from those that don't. However, to the extent certification becomes an expected feature of the product by the buyers and widespread among firms, its value to the firm, unlike the value of a brand as a source of competitive advantage, is diminished. Certification then becomes a competitive necessity rather than a source of competitive advantage.

PROMOTION STRATEGY

In this section we discuss promotion strategy of high-tech firms in the context of the appropriability problem, with particular reference to the pharmaceutical industry. A *strategic view* of promotion in high-tech firms suggests that promotion is a necessary complement to innovation (people don't buy new products if they are not aware of them) and an important instrument of modern marketing strategy by which firms mitigate the problem of appropriability. Creating and sustaining a brand name for the innovation is an essential element of this strategy. By contrast, the standard textbook treatment of promotion strategy focuses mainly on the *optimal allocation* of a firm's resources to various forms of promotion, and in measurement of their effectiveness in terms of achieving sales, share, and profitability objectives. Our focus here is the strategic role of promotion in the innovation process.

A STRATEGIC VIEW OF PROMOTION

The conceptual framework for this view, which was presented by Vinod and Rao (2000), goes like this. For firms that invest heavily in R&D, the dominant source of differentiation is in the form of new and improved products. The relevant market structure in which such firms operate is monopolistic competition, discussed in chapter 3. In his seminal work, *The Theory of Monopolistic Competition*, Chamberlin (1933, p. 56) argued that differentiation, "real or fancied, as long as it is of any importance whatsoever to buyers, and leads to preference for one variety of product over another" is what leads to "monopolistic competition." That is, neither the force of monopoly nor the force of competition excludes the other. The notion that differentiation is the key to achieving sustainable competitive advantage runs through the works of Porter (1985), Ghemawat (1986), and other scholars of competitive strategy. As Czepiel (1992, p. 50) put it, "differentiation works to provide a sustainable competitive advantage, because in delivering unique benefits valued by the market, the firm effectively removes itself from direct competition." That is, differentiation allows the firm to achieve a degree of monopoly power to avoid direct *price* competition.

However, as we discussed in chapter 1, competitive advantage gained from innovation is often not sustainable—without investments in complementary marketing assets in promotion, distribution and the like—because of imitation by competitors. As Teece (1987) noted, it is not uncommon for competitors to have profited more from innovation than the first firm that commercialized it. Put differently, the returns to investments in R&D would be reduced without the complementary investment in promotion and other marketing assets. Still, the relationship between R&D and promotion is not so straightforward, even within the same industry. It depends on the nature of the appropriability regime faced by the innovating firm as well as a variety of differentiation strategies available to the firms in the same industry. This is discussed below, taking the case of pharmaceutical industry, which offers a rich variety of differentiation strategies based on R&D and promotional effort.

DIFFERENTIATION STRATEGIES: THE CASE OF PHARMACEUTICALS

The R&D and promotional intensity matrix in Figure 6.6 describes a stylized version of four competitive differentiation strategies in the pharmaceutical industry. Firms in the upper left quadrant—with high R&D as well as promotional intensity—are of particular interest to us. This is the group of firms for whom continuous innovation of new and improved prescription drugs through R&D is the dominant source of differentiation. They also engage in high levels of complementary promotional activity in the form of detailing (i.e., personal selling to physicians, which is a large fraction of total promotion expenditures), sales promotion to physicians (e.g., free samples), direct-to-consumer advertising, creating and sustaining brand name, and so on. As discussed in chapter 3 (see "The Case of Temporary Monopoly in Pharmaceuticals"), they have legal protection from imitation during the period of *effective* patent life of typically about ten years and face intense price competition from producers of generic drugs—domestic and foreign—after the patent expires (i.e., from firms in the upper and lower right quadrants) and some competition from other patented and branded drugs during the patent life as well.

At this point it is important to note that complementary promotion is just one means by which firms producing patented and branded prescription drugs realize higher returns on R&D through greater appropriability. They also include preemptive pricing strategies prior to patent expiration to discourage post-patent entry by generics; marketing or licensing a lower-priced generic aimed at a price-sensitive segment (e.g., hospitals) while maintaining or even raising prices in the inelastic branded drug segment where prescription behavior of physicians and third-party payments favor such a strategy; patenting over-the-counter drugs; and brand extensions. Moreover, microsegmentation and targeting are particularly attractive sources of competitive differentiation in the pharmaceutical sector. This is because factors such as unique and often complex needs of the patient-consumer, the prescription behavior of doctors, and promotional efforts aimed at building brand name limit substitutability of drugs.

Promotional Strategy over Patent Life

Given the above description of competitive conditions in the market for patented prescription drugs, Figure 6.7 presents a stylized view of the promotional strategy of the firms described in the upper-left quadrant of Figure 6.6, over the life of the patent and beyond. This view is based on empirical evidence produced by classic studies on the pharmaceutical sector in the 1990s (Caves, Whinston, and Hurwitz 1991; Office of Technology Assessment 1993). The stylized graph in Figure 6.7 portrays promotional intensity (the ratio of all promotional expenditures to sales) and market share in dollars as well as units over the life of a patented and branded prescription drug. As might be expected, promotional intensity starts at 100 percent in the year of introduction, drops to about 40 percent by the third year, and stays at that level through the life of

Figure 6.6 **R&D and Promotional Intensity Matrix**

Promotional Intensity	R&D Intensity: High	R&D Intensity: Low
High	Product differentiation through R&D dominates. Promotion is complementary to R&D. Example: Traditional pharmaceutical firms and producers of patented and branded drugs, such as GlaxoSmithKlein	Promotion is the dominant differentiating factor. Example: Producers of non-patented and branded generic drugs, such as Mylan Inc.
Low	R&D is the dominant source, if not the sole source, of differentiation. Scientific discoveries through R&D are often the firm's raison d'être. Example: Biotech firms like Genentech and Chiron	Neither R&D nor promotion is a significant source of differentiation. Strategies like cost leadership and selective targeting are used to gain competitive advantage. Example: Producers of commodity generics

Source: Adapted from Vinod and Rao (2000).

Figure 6.7 **Stylized View of Innovators' Market Share and Promotional Intensity over Patent Life**

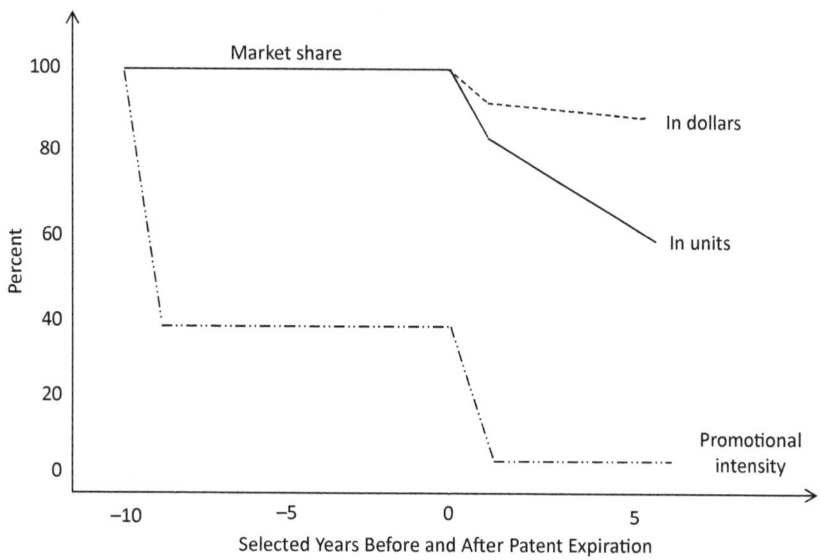

Source: Adapted from the Office of Technology Assessment (1993). Appendix F, Table F, p. 297, and Appendix G, pp. 303 and 304.

the patent. Once the patent expires, promotional intensity drops to about 7 percent in year 1. Note that the level of promotional expenditures—although low relative to sales—could be very high because of the large base of sales one would expect by post-patent year 1. With respect to market share, the figure assumes no competition during the patent life, so the market share of sales measured in dollars, as well as in units, is 100 percent. However, after the patent expires, market share in units drops sharply while the drop in dollar share is modest, indicating a counterintuitive price *increase* (in the inelastic or less price-sensitive segments of the market) in the face of cut-throat price competition from the producers of generic drugs.

The above description of promotional intensity and market share over the life of the patent indicates that the promotional strategy of firms specializing in the production and marketing of patented and branded drugs is shaped by a combination of strong appropriability regime during the effective life of the patent and expected cut-throat post-patent price competition from the generics. Building a strong brand name during the patent life is central to the promotional strategy. Indeed, the extraordinarily high R&D intensity maintained during the patent life is intended to build a strong brand name that would counter intense post-patent price competition. It is the strong brand name that makes the demand curve less sensitive to price in the consumer segment and enables firms to even raise the price in the inelastic segment, while at the same time matching the price cuts by the generic firms in the more price-sensitive institutional segment (see discussion in chapter 3). Thus, innovating firms view at least part of the heavy promotional expenditures during the patent life as *complementary investment* to R&D when competitive threat is small and appropriability strong. Vinod and Rao (2000) tested this proposition econometrically—by means of a cross-section sample of data for 53 firms and confirmed the complementary relationship between promotion and R&D. The data also indicate that among the 53 "survivor firms" that comprised the sample (i.e., those firms that were among the top 100 consistently every year through 1994), virtually all firms were in the upper left quadrant in Figure 6.6, suggesting that survivor firms recognize that in order to survive, they must not overemphasize R&D at the expense of promotion, and vice versa. We argue that the strategic view of promotion as complementary to R&D investment—whose returns are spread over the life of a product—holds for all high-tech products, albeit in a different form, depending on the unique characteristics of the consumers, the nature of technology, competitive conditions, and the like.

SUMMARY

Product strategy is central to creating and sustaining competitive advantage in all firms—high-tech and low-tech alike. In marketing, nothing really happens until someone comes up with an idea for a product that satisfies an unmet need of a market segment or segments and takes it at least through successful concept testing. Product testing follows, involving small-scale production and use by small samples of consumers. This in turn is followed by real-life market tests with all elements of the marketing program in full play. Our objective in this chapter has been to address product and promotion strategy issues that are particularly important to high-tech firms.

They include co-alignment of technology-based product concept with segmentation, issues concerning design and standards, diffusion of innovations, branding innovations, marketing of technologies, open innovation, and platform strategy.

Technology plays a central differentiating role in product strategy, and product benefits need to be co-aligned with technology for successful segmentation, targeting, and positioning strategies. A firm with dominant design can lock out its competitors and set back future generations of products in the category. Appropriability is greater when a firm's design becomes dominant in its industry. However, it must possess, among other things, strong IPRs, first-mover advantages, strength in complements, and reputation and brand name to prevail in the competitive battle over designs and standards.

Understanding diffusion of innovations is important for developing strategy and ultimately for the success of new product introductions. Diffusion is the process by which individuals and firms adopt a new/improved product, service, or technology. Applications of diffusion models involving strategic decisions include pre-launch (e.g., forecasting of new product sales), post-launch decisions (e.g., timing of successive generations of a product), and impact of pricing and other marketing program elements on diffusion. The rate at which products are diffused varies widely across product categories.

In his classic book *Diffusion of Innovation* (1995), Everett Rogers identified five characteristics that influence the rate at which innovations are adopted by consumers: *the relative advantage* of the new product (i.e., to the one it supersedes); *compatibility* with the needs of potential adopters; *complexity* (e.g., ease of use); *trialability* (i.e. can it be tested?); and *observability* (are the results easily observable?). Rogers also provided a methodology to categorize consumers into five groups on the basis of how quickly they adopt innovations: *innovators* (2.5 percent), *early adopters* (13.5 percent), *early majority* (34 percent), *late majority* (34 percent), and *laggards* (16 percent). He argued that the adoption curve should have normal distribution, shown in Figure 6.2, and the plot of cumulative number of adopters should have an S-shaped curve, shown in Figure 6.3.

The Rogers model has been criticized for lack of predictability. This is because individuals tend to be innovators in certain areas and laggards in others. The assumption of a normally distributed adoption curve also has been criticized. As a result, the Bass model, which overcomes the criticisms of the Rogers model, has become the standard diffusion model for marketing scholars and practitioners alike. The Bass model assumes that there are two groups that comprise potential adopters of an innovation and they are segmented according to the type of communication by which they are influenced. One group is influenced by mass media (external influence) and labeled as "innovators" and the other by the word of mouth (internal influence) labeled as "imitators." Imitators are influenced by previous buyers prior to purchase, whereas innovators are not. Empirical evidence on diffusion models suggests that the influence of word-of-mouth communication (internal influence) was found to be much more important than mass-media communication (external influence).

An economist's perspective on factors influencing the rate of diffusion would emphasize network effects—which arise when the value of some new product or service to each consumer increases as more and more consumers adopt it—including information and uncertainty, and industry environment in terms of market structure and regulation. Innovation by users and the role of lead users in the diffusion process has been the subject of a small but growing body of literature. The importance of lead users for the diffusion process arises from the fact that they are in need of solutions before many other users and they expect to receive relatively high benefits from obtaining a solution.

Many high-tech firms have an opportunity to gain competitive advantage through platform strategy, a key element of product strategy in many firms. At the simplest level, a platform strategy involves using a common base of components or technology—"a platform"—that enables firms to implement a competitive differentiation strategy (i.e., targeting multiple segments) without losing the benefits of scale economies. Often, platform strategy requires collaboration with producers of complementary products and opening one's otherwise proprietary technology to complementors. An industry platform with network effects leads to more users adopting the platform, which in turn leads to more users and complementors creating bandwagon effects. Amazon's evolution from online retailer of books to creating an Amazon platform, which provides e-services to other retailers, is a case in point.

Successful branding of a product or service ranks among the highest sources of sustainable competitive advantage. Branding is a critical component of product as well as promotion strategy. The concept of brand equity has to do with a set of assets and liabilities associated with a brand and the differential effect of brand as providing value to the customer and the firm. *Brand loyalty, name awareness, perceived quality, brand associations, and other proprietary assets* (e.g., trademark) comprise the five broad categories of assets and liabilities associated with a brand, as identified by Aaker. Branding innovations help a firm appropriate the economic benefits from new products, which could otherwise become commodities. Branding also aids in communicating complex features of a product; helps create a subcategory (e.g., iPod as a subcategory of portable MP3 players); and helps create and sustain the reputation of the firm for innovativeness and enhances its overall corporate brand. Note that quality certification such as ISO 9001 for software plays a significant role in enhancing the reputation of the firm as well as the brand name. Reputation and brand, although related, are not one and the same. Whereas brand is a "customer-centric" concept that focuses on product or service to the customer, reputation is a broader "company-centric" concept with focus on the credibility of an organization. Producers of complementary ingredients—materials, components, and parts—engage in ingredient branding. The strategy of ingredient branding involves taking a B2B product and turns it into a B2C product by branding the component (e.g., the "Intel Inside" label on computers).

The growing market for technologies, as opposed to products, presents special problems for strategy because technologies suffer from the appropriability problem to a greater extent than physical products. There is a strong element of moral hazard in transactions involving technologies. That is, often the seller has to disclose important information about the technology before the purchase and once it is disclosed, the buyer, in the extreme case, has no obligation to buy the technology. Market boundaries are difficult to define and shift often, making segmentation and targeting difficult. Distribution of technology is similar to communication of valuable information, which is difficult to protect from leakage. Licensing is the chief means by which technology is transacted. Ingredient branding offers one of the most promising avenues for branding technologies (e.g., the Android operating system).

A strategic view of promotion in the high-tech context would focus on complementarity between R&D and promotional efforts of a firm. This is because competitive advantage gained from R&D investments that produce new products is often not sustainable without complementary investments in marketing assets—promotion, branding, distribution. Promotion ranks among the top of marketing assets because people do not buy a new high-tech product, much less any new product they are not aware of. Promotion, along with other marketing assets, increases the appropriability of innovation and hence returns on R&D investments by shielding the firm from the threat of commoditization through imitation and price competition. Product and promotional strategies followed by pharmaceutical firms specializing in the production and marketing of patented and branded prescription drugs (described in this chapter) provide a detailed example of this view.

KEY TERMS

brand equity
diffusion of innovation
dominant design
ingredient branding
lead users
open services innovation
platform strategy
product concepts: *generic, expected, augmented,* and *potential*
promotional intensity
R&D intensity
reputation

DISCUSSION QUESTIONS

1. In what respects is product strategy similar and different in high-tech firms verses other firms?
2. Explain key product concepts and their relationship to segmentation.
3. What are the critical assets a firm must possess to win the battle over standards?
4. Explain the relevance of the following to the emergence of dominant design and its timing:
   (a) *Appropriability*
   (b) *Network effects*
   (c) Network of firms in the *value net*
   (d) Type of *standards-setting process*
   (e) *R&D intensity*
5. Explain the key determinants of diffusion and the adopter categories identified by Rogers. What are the missing determinants from an economist's perspective?
6. What are the criticisms leveled against Rogers's diffusion model? Explain the Bass model of diffusion and how it is an improvement upon the Rogers model.

7. Explain the notion of innovation by users and the role of lead users in the diffusion process.
8. Explain the role of network effects in the diffusion process.
9. Explain the relationship between platform strategy and network effects.
10. Explain the case for open services innovation and its relationship to platform strategy.
11. What are the benefits of branding innovations to the firm?
12. Explain the conditions for success in ingredient branding.
13. Explain the difference between brand and reputation.
14. In what respects is the market for technologies similar and different from the market for products?
15. Explain the significance of the concept of disruptive technology to product strategy.

REFERENCES

Aaker, David. 1991. *Managing brand equity.* New York: Free Press.

———. 1996. *Building strong brands.* New York: Free Press.

———. 2000. *Brand advertising.* New York: Free Press.

———. 2007. Innovation: Brand it or lose it. *California Management Review* 50 (1) (Fall): 8–24.

Aaker, David A., and Robert Jacobson. 2001. The value relevance of brand attitude in high-technology markets. *Journal of Marketing Research* 38 (4): 485–493.

Arora, Ashish, and Jai Asundi. 1999. Quality certification and the economics of contract software development: A study of the Indian software industry. NBER Working Paper No. 7260. Cambridge, MA: National Bureau of Economic Research.

Aurora, A., A. Fosfuri, and A. Gambardella. 2001. *Markets for technology.* Cambridge, MA: MIT Press.

Bass, Frank M. 1969. A new product growth model for consumer durables. *Management Science* 15 (5) (January): 215–227.

Baumol, W.J. 2002. *The free-market innovation machine.* Princeton, NJ: Princeton University Press.

Berndt, Ernst R., Robert S. Pindyck, and Pierre Azoulay. 1999. Network effects and diffusion in pharmaceutical markets: Antiulcer drugs. NBER Working Paper No. 7024. Cambridge, MA: National Bureau of Economic Research.

Bidault, Francis, and William A. Fischer. 1994. Technology transactions: Networks over markets. *R&D Management* 24 (4): 373–386.

Bishop, Martin. 2011. The rise of Android, another killer ingredient brand. *What's Up Below Deck Blog.* Available at http://www.landor.com/index.cfm?do=thinking.blog&post_id=25360&bhcp=1.

Carpenter, Gregory S., Rashi Glazer, and Kent Nakamoto. 1994. Meaningful brands from meaningless differentiation: The dependence on irrelevant attributes. *Journal of Marketing Research* 31 (3): 339–350.

Caves, Richard E., Michael D. Whinston, and Mark A. Hurwitz. 1991. Patent expiration, entry and competition in the U.S. pharmaceutical industry, pp. 1–48. *Bookings Papers: Microeconomics.* Washington, DC: Brookings Institution Press.

Chesbrough, Henry. 2011. The case for open services innovation: The commodity trap. *California Management Review* 53 (3) (Spring): 5–20.

Christensen, C.M. 1997. The innovator's dilemma: How new technologies cause great firms to fail. Boston: Harvard Business School Press.

Christensen, C.M., and M.E. Raynor. 2003. *The innovator's solution: Creating and sustaining successful growth.* Boston: Harvard Business School Press.

Christensen, J.F. 2006. Whither core competency for the large corporation in an open innovation world? In *Open innovation: Researching a new paradigm*, ed. H. Chesbrough, pp. 35–61. Oxford: Oxford University Press.

Christodoulides, George, and Leslie de Chernatony. 2010. "Consumer based brand equity conceptualization and measurement: A literature review." *International Journal of Market Research* 52 (1): 43–66.

Cooper, A.C., and D. Schendel. 1992. How established firms respond to threatening technologies. *Academy of Management Executive* 6 (2): 55–70.

Cusumano, Michael. 2010. Technology strategy and management: The evolution of platform thinking. *Communications of the ACM* 53 (1): 32–34. Available at http://ebusiness.mit.edu/research/papers/2010.01_Cusumano_Technology%.20Strategy%20and%20Management_272.pdf.

Czepiel, John A. 1992. *Competitive Marketing Strategy.* Englewood Cliffs, NJ: Prentice Hall.

Enos, J.L. 1962. *The rate and direction of inventive activity: Economic and social factors*, ed. R.R. Nelson, pp. 299–322. Princeton, NJ: Princeton University Press.

Etterson, Richard, and Jonathan Knowles. 2008. Don't confuse reputation with brand. *MIT Sloan Management Review* 49 (2) (Winter): 19–21.

Federal Reserve Bank of Dallas. 1996. *The economy at light speed: Technology and growth in the information age—and beyond.* Annual Report. Exhibit D. Available at http://dallasfed.org/assets/documents/fed/annual/1999/ar96.pdf.

Franke, N., and E. von Hippel. 2003a. Finding commercially attractive user innovations. Working Paper. MIT Sloan School of Management.

———. 2003b. Satisfying heterogeneous user needs via innovation toolkits: The case of Apache security software. *Research Policy* 32 (7): 1199–1215.

Freeman, C. 1968. Chemical process plant: Innovation and the world market. *National Institute Economic Review* 45 (August): 29–57.

Gawer, Annabelle, and Michael A. Cusumano. 2008. How companies become platform leaders. *MIT Sloan Management Review* 49 (2) (Winter): 28–55.

Ghauri, Pervez N., and P.M. Rao. 2009. Intellectual property, pharmaceutical MNEs and the developing world. *Journal of World Business* 44 (2): 206–215.

Ghemawat, Pankaj. 1986. Sustainable competitive advantage. *Harvard Business Review* 64 (5) (September-October): 53–59.

Glazer, Rashi. 1991. Marketing in an information intensive environment: Strategic implications of knowledge as an asset. *Journal of Marketing* 55 (October): 1–19.

Griliches, Zvi. 1957. Hybrid corn: An exploration in the economics of technological change. *Econometrica* 25 (4) (October): 501–522.

Hall, Bronwyn H. 2004. Innovation and diffusion. NBER Working Paper Series, Working Paper 20212 (January). Cambridge, MA: National Bureau of Economic Research.

Hoover International. 2011. Available at http://subscriber.hoovers.com.

Interbrand. 2011. Best Global Brands 2010. Available at http://www.Interbrand.com/en/best-global-brands.

Isckia, Thierry, and Dennis Lescop. 2009. Open innovation within business ecosystems: A tale from Amazon.com. *Communications & Strategies* 74 (2): 37–54.

Jahn, Gabriele, Matthias Schramm, and Achim Spiller. 2005. The reliability of certification: Quality labels as a consumer policy tool. *Journal of Consumer Policy* 28 (1): 53–73.

Johanson, J., and L.G. Mattson. 1987. Interorganizational relations in industrial systems: A network approach compared with the transaction cost approach. *International Studies of Management Organization* 17 (1): 34–38.

Keller, K.L. 2003. *Strategic brand management: Building, measuring, and managing brand equity*, 2d ed. Upper Saddle River, NJ: Prentice Hall.

Keller, Kevin Lane. 2008. *Strategic brand management.* Upper Saddle River, NJ: Pearson Prentice Hall.

Keller, K.L., and D.A. Aaker. 1998. Corporate-level marketing: The impact of credibility on a company's brand extensions. *Corporation Reputation Rev.* 1 (August): 356–378.

Keller, Kevin Lane, and Donald R. Lehmann. 2006. Brands and branding: Research findings and future priorities. *Marketing Science* 25 (6): 740–759.

Kotabe, Masaaki, Arvind Sahay, and Preet S. Aulakh. 1996. Emerging role of technology licensing in the development of global product strategy: Conceptual framework and research propositions. *Journal of Marketing* 60 (January): 73–88.

Kotler, Phillip. 1991. *Principles of marketing.* 7th ed. Upper Saddle River, NJ: Prentice Hall.

Kotler, Phillip, and Waldemar Pfoertsch. 2010. *Ingredient branding: Making the invisible visible.* Heidelberg: Springer.

Lehmann, Donald R., and Russell S. Winer. 2005. *Product Management.* 5th ed. New York: McGraw-Hill.

Levitt, Theodore. 1980. Marketing success through differentiation—of anything. *Harvard Business Review* (January–February): 1–10.

Lilien, G.L., P.D. Morrison, K. Searls, M. Sonnack, and E. von Hippel. 2002. Performance assessment of the lead user idea-generation process for new product development. *Management Science* 48 (8): 1042–1059.

Lüthje, C. 2004. Characteristics of innovating users in a consumer goods field: An empirical study of sport-related product consumers. *Technovation* 24 (9): 638–695.

Mahajan, Vijay, Eitan Muller, and Frank M. Bass. 1990. New product diffusion models in marketing: A review and directions for research. *Journal of Marketing* 54 (January): 1–26.

Mahajan, Vijay, Eitan Muller, and Rajendra K. Srivastava. 1990. Determination of adopter categories by using innovation diffusion models. *Journal of Marketing Research* 27 (February): 37–50.

Mahajan, Vijay, Eitan Muller, and Yoram Wind. 2000. New-product diffusion models: From theory to practice. In *New-product diffusion models.* London: Springer Science.

Mergent Online. 2011. Available at http://www.mergentonline.com.

Morgan, Robert M., and Shelby D. Hunt. 1994. The commitment—Trust theory of relationship marketing. *Journal of Marketing* 58: 20–38.

Morrison, P.D., J.H. Roberts, and D.F. Midgley. 2004. The nature of lead users and measurement of leading edge status. *Research Policy* 33 (2): 351–362.

Noteboom, Bart. 1996. Trust, opportunism and governance: A process control model. *Organization Studies* 17 (6): 985–1010.

Office of Technology Assessment. 1993. *Pharmaceutical R&D: Costs, risks, and rewards*, pp. 295–304. Washington, DC: U.S. Government Printing Office.

Porter, Michael E. 1985. *Competitive advantage: Creating and sustaining superior performance.* New York: Free Press.

Pisano, Gary. 1997. *The development factory.* Boston: Harvard Business School Press.

———. 2006. Profiting from intellectual property revolution. *Research Policy* 35: 1122–1130.

Rao, P.M., and Joseph A. Klein. 2008. Innovation by users and competitive strategy. Paper presented at the American Society for Competitiveness, 19th annual conference, November 6–8, Orlando.

Rao, P.M., V. Vemuri, and P. Galvin. 2004. The changing technological profile of the leading ICT firms: Evidence from U.S. patent data, 1981–2000. *Industry and Innovation* 11 (4): 353–372.

Roberts, James W. 2011. Can warranties substitute for revolution? *American Economic Journal: Microeconomics* 3 (3) (August): 69–85.

Rogers, Everett M. 1962. *Diffusion of innovations.* New York: Free Press.

———. 1995. *Diffusion of innovations.* 4th ed. New York: Free Press.

———. 2003. *Diffusion of innovations.* 5th ed. New York: Free Press.

Sanchez, Ron. 1999. Modular architectures in the marketing process. *Journal of Marketing* (Special Issue) 63: 92–111.

Sawhney, Mohanbir. 2005. Branding in technological markets. In *Kellogg on Branding*, ed. Alice M. Tybout and Tim Calkins, pp. 201–225. New York: John Wiley & Sons.

Shah, S. 2000. Sources and patterns of innovation in a consumer products field. Working Paper, MIT Sloan School of Management.

Shapiro, Carl, and R. Hal Varian. 1999. The art of standard wars. *California Management Review* ABI/INFORM Global, 41 (2): 8–32.

Simon, Carol J., and Mary W. Sullivan. 1993. The measurement and determinants of brand equity: A financial approach. *Marketing Science* 12 (1) (Winter): 28–52.

Srinivasan, Raj, Gary L. Lilien, and Arvind Rangaswamy. 2006. The emergence of dominant designs. *Journal of Marketing* 70 (2) (April): 1–17.

Sultan, Fareena, John U. Farley, and Donald R. Lehmann. 1990. A meta-analysis of applications of diffusion models. *Journal of Marketing Research* 27 (1) (February): 70–77.

Teece, David J. 1987. Profiting from technological innovation: Implications for integration, collaboration, licensing, and public policy. In *The Competitive Challenge*, pp. 185–219. Cambridge, MA: Ballinger.

Urban, G.L., and E. von Hippel. 1988. Lead user analyses for the development of new industrial products. *Management Science* 34 (5): 569–582.

Utterback, J.M. 1994. *Mastering the dynamics of innovation.* Boston: Harvard Business School Press.

Utterback, James, and Happy Acee. 2005. Disruptive technologies: An expanded view. *International Journal of Innovation Management* 9(1): 1–17.

Vinod, H.D., and P.M. Rao. 2000. R&D and promotion in pharmaceuticals: A conceptual framework and empirical exploration. *Journal of Marketing: Theory and Practice* 8 (4) (Fall): 10–20.

von Hippel, E. 1998. Economics of product development by users: The impact of sticky local information. *Management Science* 44 (5): 629–644.

———. 2005. *Democratizing innovation.* Cambridge, MA: MIT Press.

West, J., W. Vanhaverbeke, and H. Chesbrough. 2006. Open innovation: A research agenda. In *Open innovation: Researching a new paradigm*, ed. H. Chesbrough, pp. 285–307. Oxford: Oxford University Press.

7 Externalities and Ethics

The Social Impact of Technology and How High-Tech Firms' Responses May Intertwine with Their Marketing Strategies

Doing well is the result of doing good. That's what capitalism is all about.
—Ralph Waldo Emerson

I think many people assume, wrongly, that a company exists simply to make money. While this is an important result of a company's existence, we have to go deeper to find our real reason for being . . . A group of people get together and exist as an institution that we call a company . . . to do something worthwhile—they make a contribution to society.
—David Packard, co-founder of Hewlett-Packard

Chapter Overview

An externality is a cost or benefit that is borne or enjoyed by someone not directly involved in the production or consumption of a good or service. In chapter 1 we discussed the externalities affecting the appropriability of intangible intellectual capital, which is expensive to produce but cheap to copy, and how reliance on intellectual property law alone to protect the private return from innovations falls short. We examined how creating and sustaining a brand name and other difficult-to-imitate marketing strategies are significant means by which successful innovators protect themselves against imitation. We also discussed network externalities—the bandwagon effects that are also sometimes referred to more technically as external demand-side increasing returns or external demand-side scale economies—as a special characteristic of technological innovation.

Our purpose in this chapter is to examine another relationship between externalities and technological innovation—how the use of technology by firms and entire industries in their businesses can produce social costs such as environmental harms and social benefits such as improvements in delivery of health care to impoverished regions of the world, and how high-tech firms are dealing with these social costs and benefits.

We consider in this chapter how high-tech firms' ethical choices address such externalities beyond what the law may dictate. By ethical choices, we mean the selection of business practices that are consistent with the firm's normative standards of acceptable behavior. A firm's ethical standards are influenced by the business and cultural environment in which it operates, including the expectations of stakeholders such as its customers, employees, other firms with which it does business, the media, civil society, and public policy makers.

A firm's ethical choices affect its reputation and brand, which are generally considered related but distinct concepts as we noted in the chapter on product strategy. According to Ettenson and Knowles (2008), reputation is "a company-centric concept that focuses on the credibility and respect that an organization has among a broad set of constituencies" and brand is a "customer-centric concept that focuses on what a product, service or company has promised to its customers and what that commitment means to them" (p. 19). Where we see reputation and brand come closest together, according to Ettenson and Knowles, is in "companies that boast corporate ethics as the core of their brand positioning" (p. 20).

This chapter focuses on how technology, externalities, business ethics, and marketing strategies are intertwined in high-tech markets. We discuss and provide examples of how a firm's public demonstration of ethical behavior plays into its marketing strategies to enhance company reputation and improve customer image of its brand, behavior that would in turn help mitigate negative externalities and internalize positive externalities even in the absence of effective legal rules and enforcement mechanisms. In this context, we introduce the concept of *social innovation*, defined by Phills, Deiglmeier, and Miller of the Stanford Center for Social Innovation (2008) as follows: "a novel solution to a social problem that is more effective, efficient, sustainable, or just than existing solutions and for which the value created accrues primarily to society as a whole rather than private individuals" (Phills Jr., Deiglmeier, and Miller 2008, p. 36).

After a general discussion of these issues, we shall use two case studies to illustrate how certain externalities, and the manner in which high-tech firms deal with these externalities, affect areas of major social concern to many people—the environment and privacy. We conclude this chapter with a brief review of ethical best practices.

TECHNOLOGY AND ITS EXTERNALITIES

There is hardly an aspect of human life that is not affected by technological change. While technological change is at the heart of improved standards of living the world over, the system of incentives or disincentives underlying the creation, production, marketing, and consumption of certain technologies—as opposed to technology per se—can result in undesirable side effects that impose societal costs and useful benefits that enhance societal well-being.

Economists use the term "negative externalities" to refer to the harm to others caused by activities or behavior of an individual, a firm, an industry, or a country furthering their own private interests but not incurring the full cost of the harm suffered by others. Think of climate change and rising sea levels, for example, resulting from excessive production and use of fossil fuels. Applying technological solutions solely to create efficiencies and lower costs for the producers and/or direct users of the producers' output may literally or figuratively dump associated costs on society and the planet. When negative externalities are present, the total cost of producing a good or service, known as the social cost, exceeds the private cost alone by the amount of external cost that is imposed on the society at large.

Technology can also be applied to produce societal benefits, such as vaccines that reduce disease and save lives, or chemicals that help increase food production, reduce malnutrition, and increase life expectancy. Economists refer to such societal benefits as "positive externalities." When there are positive external economies, market prices of products and services do not fully reflect the benefits to the society. The total benefit exceeds the private benefit realized by the producers and/or direct users of the producers' output by the amount of external benefit to the society at large.

In the presence of externalities, competitive markets either produce too much of a good or service that exacts a social cost (negative externalities), or too little of a good or service for which there are social benefits relative to the economically efficient amounts (positive externalities).

It is not surprising that most technological research and development expenditures by commercial organizations are devoted to product and process innovations that have direct economic market value to them. As a result, technologies needed to produce "social goods," such as improved air quality, or that address societal impacts of collective industrial behaviors whose costs individual firms may seek to avoid because they are not fully recoverable in their prices, such as the external costs of pollution or waste disposal, are ignored or relegated to a lower order of priority. The gaps between private and public benefits and costs are shown in the Figures 7.1 and 7.2. Because the market fails to produce the economically efficient level of output in the face of externalities, such a situation is known as *market failure.*

THE LIMITS OF PUBLIC POLICY IN ADDRESSING EXTERNALITIES

Some economists would argue that externalities and market failures result from incomplete property rights or from the difficulty of enforcing property rights. Ronald Coase (1937), in particular, showed that when property laws are well defined and transaction costs are low among the affected parties (particularly, when there are few), negotiation among parties would produce economically efficient private solutions for the problem of externalities. However, when the potential social costs or social benefits produced by technological innovation are very spread out and property laws are not well defined

Figure 7.1 **Subsidy to Encourage the Production of Goods with Positive Externality**

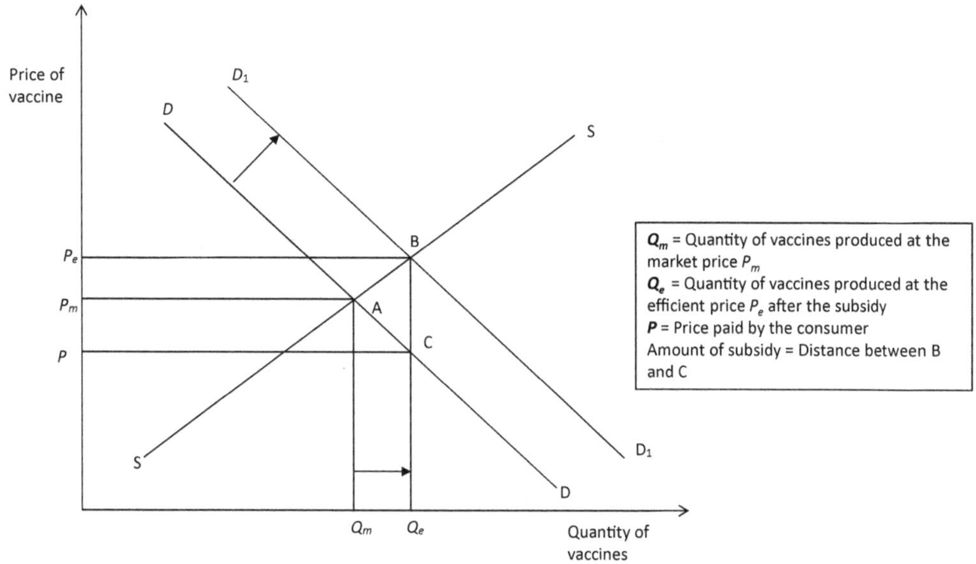

Figure 7.2 **Tax to Discourage Goods with Negative Externality**

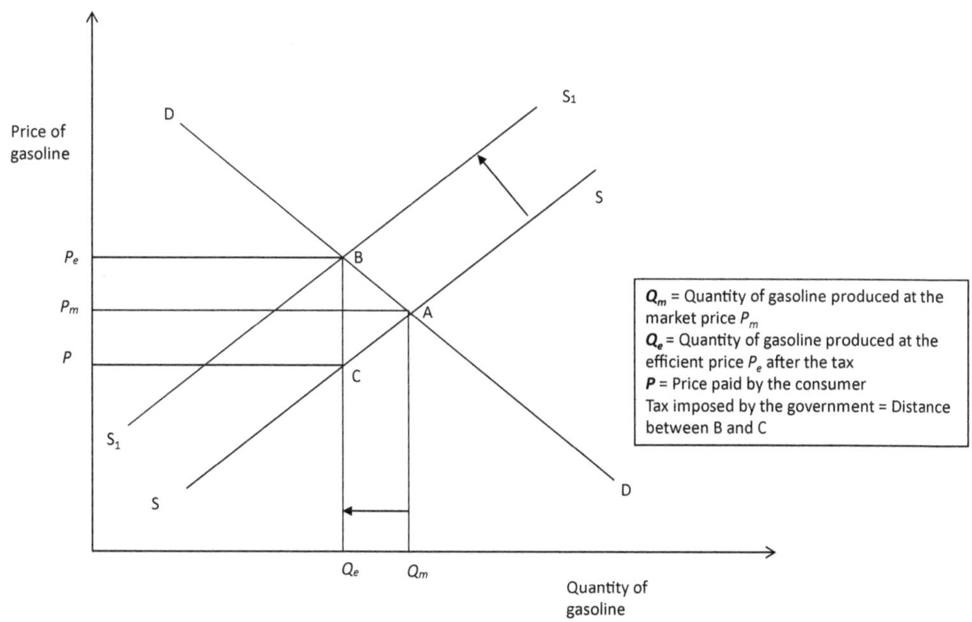

or applicable to the circumstances involving many affected parties, Coase's conditions for an economically efficient private solution are less likely to be satisfied.

Milton Friedman (1999) has argued that the largest responsibility of the company is to make profits. By making profits companies can pay employees and shareholders and contribute in that manner to the economic well-being of society. A corollary to this argument is that it is up to the government to rein in, through laws and regulations, private behaviors that have the potential for causing public harm, complemented by judicial remedies for statutory violations and common-law torts such as negligence or nuisance. Similarly, according to this line of argument, if companies are to concentrate on making profits to benefit their shareholders and employees, it is up to the government to incentivize private firms to invest in social innovations.

A variety of public policies are available to partially address the market failures associated with externalities. They range from direct controls by the government to market-based solutions involving a tax on pollution equal to damages it causes or trading pollution permits. Courts also have the tools to enforce laws addressing pollution and other negatives, to handle common-law tort claims, and to provide remedies that make the firms generating negative externalities legally liable for the damages they cause. Public policies can also play a significant role in incentivizing firms to invest in social innovations, such as R&D tax credits and other tax benefits, or public subsidies to match private investments in energy-saving technologies.

However, laws and regulations alone are inadequate to steer companies in a socially responsible direction. In our increasingly globalized economy, large multinational firms in particular can leverage their political and economic influence and find ways to conduct their far-flung operations that enable them to evade responsibility for the significant external costs they impose around them.

Even legislatively imposed market-based solutions to negative externalities, such as placing a price on the negative externality of carbon emissions through carbon taxes or the use of cap-and-trade carbon emissions policies, have encountered political obstacles to passage—in no small part because of intense lobbying by the affected industries. Moreover, enforcement of laws and regulations is difficult enough within one national jurisdiction, as technology often outpaces the legal response to its societal consequences. Enforcement becomes even more problematic when firms operate under the laws of multiple jurisdictions.

Court remedies for individuals hurt by a firm's socially irresponsible behavior are also an incomplete solution. Social damage is not easily quantifiable. Moreover, an individual claiming that he or she has been directly damaged by a firm's socially irresponsible behavior must prove that the firm violated a legally recognized standard of behavior for which the individual has standing to sue (whether embodied in a statute with a private cause of action or in a common-law tort such as negligence or nuisance) and that the damages the individual suffered were proximately caused by the firm's violation of that legal standard.

As for incentivizing firms to increase their investments in social innovations to produce positive externalities, proposals to provide direct government subsidies for research and development in such technologies as alternative energy sources to a level that can make a significant difference tend to get mired in political and budgetary disputes.

A full discussion of the impact of public policies and legal remedies on firm behavior is beyond the scope of this book. Suffice it to say that they do not provide a complete solution to the externality problem. We shall focus instead on how high-tech firms themselves decide to deal, from an integrated ethical/marketing strategy perspective, with the social costs they potentially impose on society in developing and using their technologies in their products and production processes. We shall also focus on ethical/marketing strategy incentives that drive some high-tech firms to affirmatively decide to invest in social innovation.

THE CONVERGENCE OF EXTERNALITIES, ETHICS, AND MARKETING STRATEGIES

BASELINE ETHICAL BEHAVIOR

Before discussing "corporate social responsibility" (defined below), which constitutes a sub-set of business ethical practices of particular interest in examining how high-tech companies deal with externalities as part of their marketing strategies, it is first worth noting the broader set of baseline business ethics concerns shared by high-tech and an array of other industries.

As Dr. Carl J. Schramm, President and CEO of The Ewing Marion Kauffman Foundation, one of the world's largest foundations devoted to entrepreneurship, said in summing up the importance of business ethics to society: "The costs of the transgressions of a few are imposed as a burden on the whole of society" (Schramm 2002, p. 12).

Indeed, the concept of externalities can easily be extended to the ethical domain.

The ethical choices firms make in their day-to-day business operations include how honest firms are in their financial management and communications about their financial condition, how fairly they treat their workers, how they protect the personal information of their customers and employees, how they dispose of waste materials, whether they engage in misleading advertising and communications about their products, and the tactics they use in selling their products and services.

A firm may not technically cross over the boundaries of outright illegal false advertising, bribery, fraud, or anticompetitive conduct, but might still engage in unethical behavior that has social consequences within the firm, the industry, and in the larger community.

QUESTIONABLE SALES TACTICS EXAMPLE

Sales tactics involve ethical choices that firms in all industries and markets face, no matter what products or services they sell. When does a salesperson's offer of a gift or reciprocal financial compensation to a customer as inducement to buy a product or service cross the line between socially acceptable and unacceptable behavior?

Think of a case of unethical behavior by a few key salespeople in a company having a potentially negative effect on the reputation of the entire company, and even on other businesses within and outside its industry. By way of example, questionable tactics by the sales departments of some companies that manufacture implants like heart devices and artificial joints have included trying to entice doctors to promote their products via inflated research consulting deals. A story about the questionable sales tactics of one small company was featured in the *New York Times*, recounting charges that the company had relied for its success on "developing relationships with doctors who, in turn, can influence which brand of device a patient gets." The article did not limit itself to this one company, but rather used it to raise questions about the integrity of the entire implant industry. The article said that the scores of unearthed company documents "offer a portrait of an implant industry where producers seek to influence the brand of device that patients receive long before a diagnosis" (Meier 2011).

This kind of conduct has attracted the attention of law enforcement agencies, but straddles the line between clearly illegal behavior and an unethical practice that may be technically legal but causes a negative externality.

However, while the stick of legal compliance is not a complete answer to the problem of negative externalities produced by socially harmful business practices, companies also have their reputations and protection of their brands to consider. They operate today in a 24/7 news cycle in which customer, employee, and other stakeholders' awareness of business behavior is heightened by the proliferation of information on the Internet and social media.

Does Firm Size Matter to Ethical Performance?

Large publicly traded firms in particular are paying attention to ethical issues such as human rights, labor, environment and the fight against corruption, as reported in the United Nations Global Compact's 2011 Global Compact Implementation Survey. According to the Global Compact's Executive Director: "Large corporations clearly have outperformed smaller on all scores . . . Another very important finding is by ownership, we found clearly that publicly owned companies have the highest performance record, whereas state owned are second and actually the lowest performers are the privately owned." An analysis of large and small bioscience firms' ethical behaviors came to a similar conclusion: "Large public companies were much more likely to engage in a variety of ethical initiatives than private firms" (Finegold and Moser 2006).

Against the backdrop of major corporate scandals that have destroyed companies such as Enron and WorldCom and have embarrassed companies from the pharmaceutical and energy industries to Wall Street, large publicly traded firms with a global footprint, a large, diverse customer and employee base, and high visibility in the media are aware of how much they stand to lose if they do not factor in potential societal harms when making their business decisions. For similar reasons, they may also be motivated to apply their expertise and resources to increase the social benefits even where they do not reap an immediate return on their investments.

Smaller entrepreneurial privately held firms may not scale up to the same level of resources and attention devoted to ethical issues as larger companies. However, in seeking to enhance their own reputations as they expand, and as they engage in supplier or other strategic alliances with larger, publicly traded firms that have internalized ethical business practices, the smaller entrepreneurial firms have their own incentives to act ethically. "Due to the global impact of large corporations the public's eye is primarily on large businesses; however, ethical issues arise in small businesses as well" (Cooksey and Kuchina-Musina 2010, p. 23).

Definition and Scope of "Corporate Social Responsibility"

A discussion of ethics-based business conduct, which measures an act in terms of its impact on social welfare, leads us to the concept of corporate social responsibility

(CSR). Although some academic researchers have sought to distinguish, for the purposes of their empirical studies, between the term "business ethics" as more inward-looking behaviors of individuals and groups within the firm, and the term "corporate social responsibility" as more externally focused behaviors (Baucus and Cochran 2009), we view corporate social responsibility the way businesses themselves often view it—as part and parcel of a code of ethical practices.

In a document co-authored by the Sustainability and United Nations Global Compact organizations, CSR is defined as "an approach to business that embodies transparency and ethical behavior, respect for stakeholder groups, and a commitment to add economic, social and environmental value."[1] This is a variation of what has been referred to in business literature as "triple bottom line" (people, planet, profit) (Brown, Dillard, and Marshall 2006).

According to a 2003 GlobeScan survey of consumers across G20 countries of how they would rate various industry sectors in terms of their CSR performance, the computer, telecommunications, and biotech industries ranked first, second, and fourth, respectively, of the fifteen industries included in the survey. The pharmaceutical industry ranked seventh, right behind the auto industry and just ahead of clothing/apparel. The oil/petroleum, alcohol, and tobacco industries were ranked at the bottom of the survey.

High-tech firms have maintained their high rankings. In its 2011 survey, GlobeScan found that "high-tech/computer" and telecommunications ranked first and third, respectively. The food industry moved into second place. Pharmaceuticals moved into fourth place (the biotech industry was not rated separately in the 2011 survey).

See Figures 7.3 and 7.4 for the 2003 and 2011 survey results.[2]

The fact that, as measured by GlobeScan, the high-tech sector has maintained its lead ahead of other industries in terms of its perceived CSR performance may indicate the particular attention CSR receives from high-tech firms. As we shall see, high-tech firms that are leaders in socially responsible business practices are not only demanding such behaviors from their own managers and employees. They also expect that all vendors in their supply chain incorporate the same level of social responsibility in their business practices in areas such as the environment in order to be eligible for consideration as a procurement source.

Moreover, "vanguard companies," which use their technologies and influence "to develop solutions to problems the public cares about" and practice ethical behavior above and beyond what the law requires, may incentivize competitors in the same industry to follow suit (Kanter 2009, p. 12). A reputation as a laggard company in this area may negatively affect consumers' attitudes toward a company, affecting its ability to recruit and retain skilled technical experts.

Does CSR Provide a Marketing and Competitive Edge?

According to results from a *2010 corporate social responsibility survey*, more than 75 percent of consumers indicated that it is important for companies to be socially responsible.[3] A five-country survey conducted by the market research group GfK NOP would appear to back up that finding (Piercy and Lane 2009).

Figure 7.3 **CSR Performance of Industry Sectors**

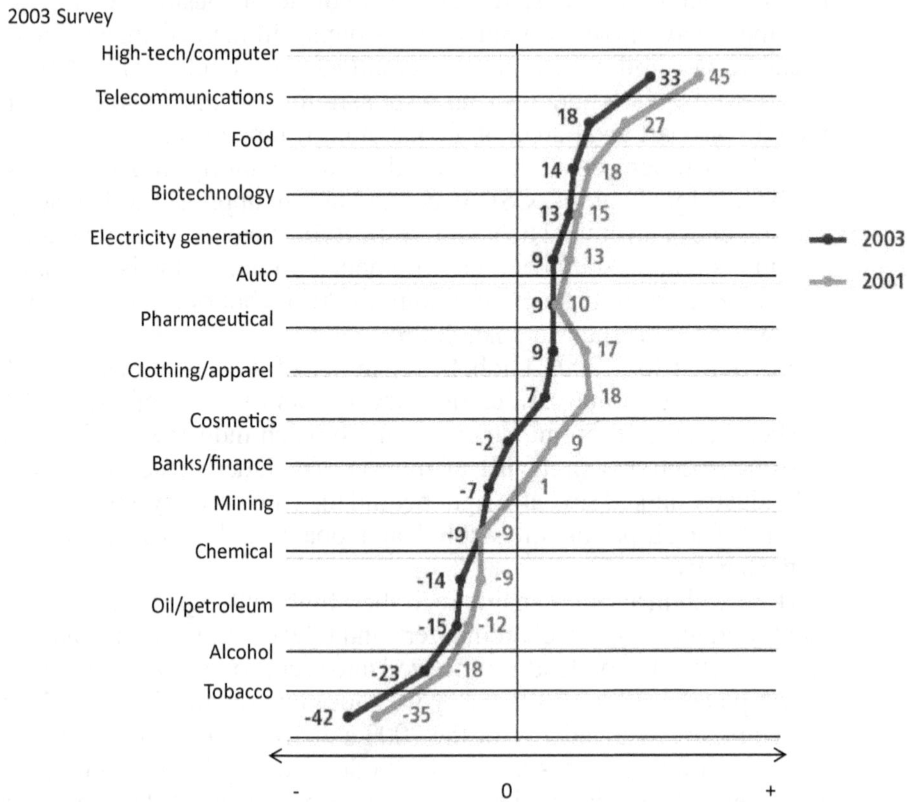

Source: Reproduced with permission of GlobeScan Inc., London, Toronto, San Francisco.

However, as the economic theory of externalities would predict, customers are not necessarily willing to pay a premium for products made by socially responsible companies. Thus, as reported by Hewlett-Packard (HP) in its global citizenship report (2010), "while more than 77 percent of consumers consider 'green' brands to be somewhat or very important when making a purchasing decision, research shows that many consumers are unwilling to pay a premium for 'green' technology" (p. 20).

O'Marah and Karofsky (2006) also found that "HP's green-friendly image" has had little immediate impact on brand enhancement that would lead mass-market buyers in the consumer and small business segments to pay a premium for HP products on account of HP's CSR activities, although they appear to have had more of an impact on the purchasing decisions of larger buyers, including CSR-oriented corporate customers and public sector customers.

The marketing literature is mixed regarding the impact of socially responsible initiatives on financial performance. Some studies canvassed by Cronin et al. (2010) focus on the additional costs, placing socially responsible firms at a financial disadvantage. Nevertheless, even if firms do not necessarily expect to be able to quickly recoup the direct costs of developing and offering their customers "green" and other

Figure 7.4 **CSR Performance of Industry Sectors, 2011 Survey**

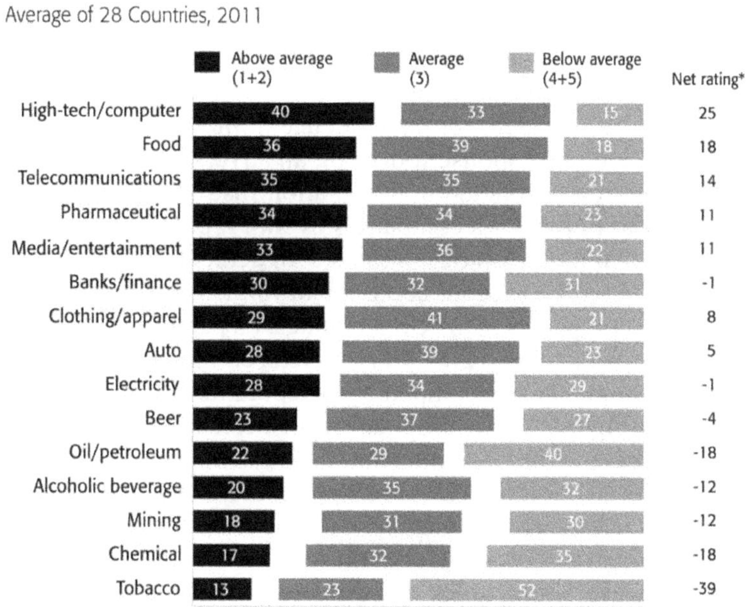

*Net ratings represent "Among the very best" and "Above average" minus "Below average" and "Among the very worst."

The white space in this chart represents "DK/NA."

Source: Reproduced with permission of GlobeScan Inc., London, Toronto, San Francisco.

socially responsible technologies, the marketing and competitive benefits that flow from being able to demonstrate "best in class" corporate ethics to their stakeholders such as customers and employees provide incentives for socially responsible behavior that help to overcome the market failure posed by externalities. The business benefits we discuss in this section include "brand insurance" and protection of reputation, employee recruitment and retention, and fostering competitive growth in emerging markets.

"Brand and Reputation Insurance" and Employee Recruitment/Retention

Customer perceptions of a firm's degree of commitment to ethical business behavior, including its level of socially responsible actions, can affect the quality of the firm's product brands over time. Werther and Chandler (2005) refer to corporate social responsibility as global brand insurance.

Minor and Morgan (2011) use the phrase "reputation insurance" in describing how "expenditures on corporate social responsibility (CSR)—both 'doing good' and 'avoiding harm'—can be thought of as an insurance premium" (p. 42). Minor and Morgan conducted a comparative study of how two medical device manufacturing firms with different approaches to the "avoiding harm" dimension of CSR fared financially. They also performed a statistical analysis of CSR and product recall event data on S&P 500 companies from 1991 to 2006. From their comparative study and statistical analysis they concluded that CSR "acts as a powerful form of reputation insurance when a firm suffers an adverse event" (p. 55). The S&P 500 analysis in particular provided "strong evidence of the link between a firm's CSR activities and its stock price response following an adverse event" (p. 55).

Research conducted by Reputation Institute (RI) found that CSR matters significantly to ratings of a company's overall reputation. In its 2008 and 2009 studies, RI found that its CSR index predicted 40 percent of overall reputation. RI also analyzed what a high CSR score could mean for a company's reputation: "Out of 200 of the largest companies in the United States, 65 percent of the general public will 'definitely recommend' the top 20 CSR American companies while only 26 percent recommend the bottom 20 CSR companies" (Reputation Institute 2010).[4]

See Figure 7.5 for Reputation Institute's Relationship Between Reputation and Support.

In its 2012 report, RI found that "more than 55 percent of consumers say that they would definitely recommend" to others the top 100 companies in reputation while fewer than 22 percent would do so for the bottom 100 companies on the list.[5]

However, companies must not only "talk the talk." They must be perceived as "walking the talk" through their actions that align with CSR, or their product brand and overall firm reputation may suffer in the marketplace (Wagner, Lutz, and Weitz 2009). Piercy and Lane (2009) observed that firms perceived as faking social responsibility will suffer consumer backlash.

Another way to look at the issues of "brand insurance" and protection of overall company reputation is from a financial risk perspective. Consider how much money marketers devote to building up positive identities for their brands and company reputations in the marketplace. Such investments by a firm in creating a positive identity can be seriously undermined by highly publicized actions of the firm that people sense are contrary to their own core values. "It may not wipe the sales out the very next day (although that has also been known) but it may still make an impact" (Baker 2006).

Baker cites Apple as an example of a high-tech firm that cultivated its brand identity and overall reputation, only to have them jeopardized by stories that its iPods were being made in sweatshops. If Apple had tried to dismiss the story out of hand or say that what its overseas suppliers did was not its business, it risked becoming known as a company that does not care about basic human rights in producing its products, and there could have been serious repercussions to its forward-looking image. However, as Baker reports, "that was not Apple's response" (Baker 2006). Instead, Apple has taken steps to investigate the labor practices of its overseas suppliers' labor practices and to require that they take remedial steps.[6]

Apparently, Apple's corrective actions have helped preserve its positive reputation in the public's eyes. According to a 2011 consumer study by the Reputation Institute, Apple ranked number 2 (right behind number 1 Google) on its list of the world's most reputable companies. See Figures 7.6 and 7.7 for the "RepTrak™ Model–Global

Figure 7.5 **Relationship Between Reputation and Support**

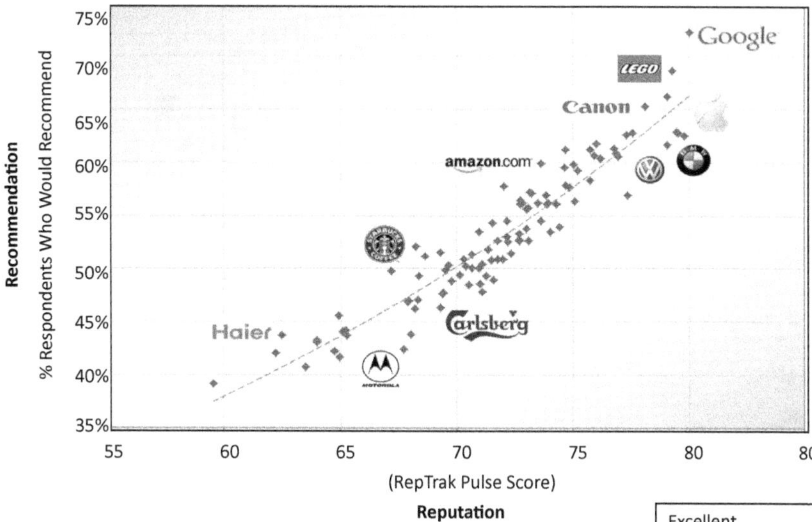

Excellent	80+
Strong/Robust	70–79
Average/Moderate	60– 69
Weak/Vulnerable	40– 59
Poor/Bottom Tier	Below 40

Source: Reproduced with permission of the Reputation Institute. RepTrak is a registered trademark of the Reputation Institute. © Reputation Institute, all rights reserved.

Figure 7.6 **The RepTrak Model on 2011 Global Drivers of Reputation**

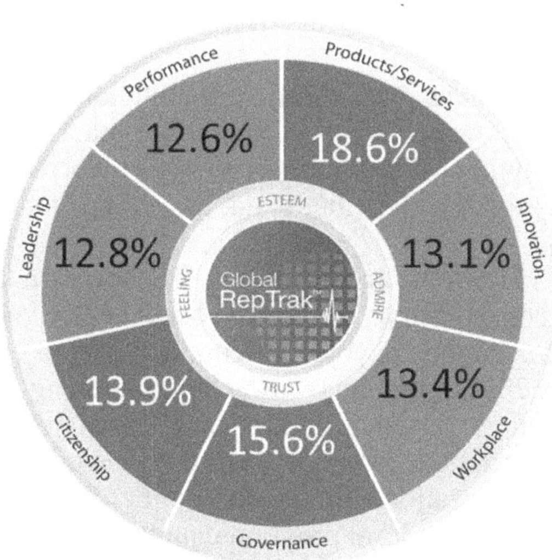

Factor Adjusted Regression
$N = 18{,}000$
$Adj - R^2 = 0.692$

Product/Services: Company offers high-quality products and services—it offers excellent products and reliable services

Innovation: Company is an innovative company—it makes or sells innovative products or innovates in the way it does business

Workplace: Company is an appealing place to work—it treats its employees well

Governance: Company is a responsibly run company—it behaves ethically and is open and transparent in its business dealings

Citizenship: Company is a good corporate citizen—it supports goo causes and protects the environment

Leadership: Company is a company with strong leadership—it has visible leaders and is managed effectively

Performance: Company is a high-performance company—it delivers good financial results

Source: Reproduced with permission of the Reputation Institute. RepTrak is a registered trademark of Reputation Institute. © Reputation Institute, all rights reserved.

Figure 7.7 **Companies with the Best Reputation in 2011: Within the 7 Dimensions of Reputation**

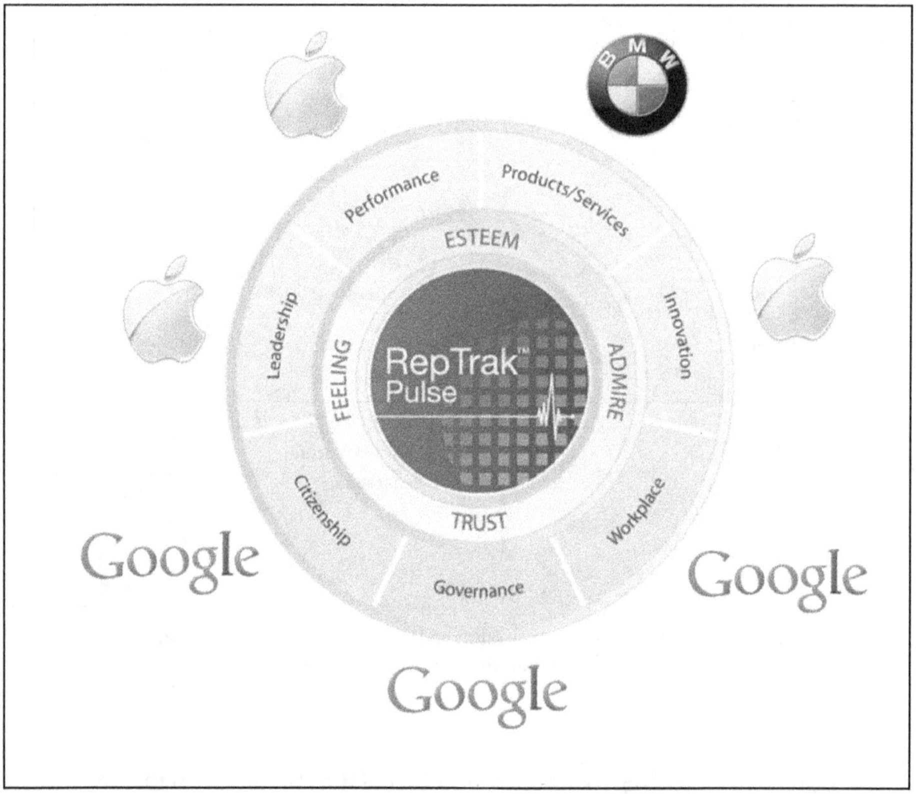

Source: Reproduced with permission of the Reputation Institute. RepTrak is a registered trademark of Reputation Institute. © Reputation Institute, all rights reserved.

Drivers of Reputation" and "Companies with the Best Reputation in 2011: Within the 7 Dimensions of Reputation," respectively. It should be noted, however, that Apple and Google dropped to number 5 and number 6 respectively in the Global RepTrak™ 100 for 2012. Microsoft, on the other hand, moved up from number 11 to 7 in the 2012 list.[7]

High-tech companies such as Hewlett-Packard (which ranked 17 on the 2011 RI list) appear willing to invest in concrete steps to help associate themselves with CSR in the public's mind, even if the economic payoff is not immediate. For example, HP was an organizing partner of the 2011 Social Innovation Summit (along with Abbott, Intel, and Microsoft), which brought together leaders from the private and public sectors at United Nations headquarters in June 2011. HP used the occasion to issue a press release announcing its collaboration with the nonprofit organization Positive Innovation for the Next Generation. The joint program equipped health-care workers in Botswana with HP Palm Pre 2 smartphones to collect malaria data, notify the Ministry of Health about outbreaks, and tag both data and disease surveillance information with GPS coordinates. The press release was carried on prominent financial

Web sites such as The Street.com and Nasdaq.com, as well as on a corporate social responsibility site for Africa that displayed HP's logo.[8]

Repeated positive associations like this over time can help protect a firm's product branding and overall reputation in the event that it should take a hit from an embarrassing incident sometime in the future. In other words, high-tech firms' investments in CSR can provide them with brand and reputation insurance.

A firm's commitment to CSR may also affect its perception in the marketplace for highly skilled employees, which is especially important to high-tech companies. As Bhattacharya, Sen, and Korschun (2008) point out, "There is growing evidence that a company's corporate social responsibility activities comprise a legitimate, compelling and increasingly important way to attract and retain good employees" (p. 37). They suggest ways that managers can use employee involvement in CSR initiatives as a way to better connect with and motivate their employees by allowing them to serve "as the primary enactors of social responsibility programs" who can "take the lead in designing and implementing CSR programs" (p. 44).

For high-tech firms facing shortages of highly trained informational technology professionals, offering such professionals the opportunity to apply their skills and experience to socially responsible projects has proven to be a valuable employee recruiting and retention tool. HP Canada, for example, conducted an employee survey that specifically asked if Hewlett-Packard's environmental programs made employees more likely to work/stay at HP. The results found that nearly 60 percent of responding employees said yes.[9]

PARTNERSHIPS, COMPETITIVE CONTEXT, AND SHARED VALUE

In a number of cases, high-tech firms choose to partner with government agencies, charitable organizations, and/or nongovernmental organizations (NGOs) in order to leverage the relationships, prestige, and infrastructure of such noncommercial entities. Cronin et al. (2010) refer to these partnerships as "social alliances."

We have already mentioned Hewlett-Packard's collaboration with the nonprofit organization Positive Innovation for the Next Generation. Another example is General Electric, which is featured on the Environmental Defense Fund (EDF) Web site for its joint initiative with the environmental group in demonstrating energy-efficiency opportunities for cities, universities, hospitals, and companies. GE's logo appears on one of EDF's Web pages.[10]

Moreover, according to Porter and Kramer (2002), such partnerships also help improve firms' "competitive context" by contributing to the betterment of the business environment in the locations where they operate. They mention programs to improve education of the local workforce as one means to improve a firm's "potential competitiveness," citing as an example Cisco's Networking Academy to train and certify secondary and post-secondary students in network administration. Cisco saw a gap that needed to be filled, benefiting regions of the United States as well as developing countries where networking skills required for knowledge-based jobs were scarce; in addition, Cisco built increased demand for its products among customers who value access to well-trained network administrators (Porter and Kramer 2002).

In a more recent article, Porter and Kramer (2011) discuss their related concept of "shared value," which they offer as an alternative way of looking at social impacts other than strictly in terms of externalities. Their theory of "shared value" postulates that addressing societal needs can actually create new markets and that social harms can create "internal costs" for firms such as wasted energy and an inadequately educated local workforce. Generating value or benefits that are shared with society, Porter and Kramer argue, can lead firms to develop and use "new technologies, operating methods, and management approaches—and as a result, increase their productivity and expand their markets" (Porter and Kramer 2011, p. 5). They cite Google, IBM, Intel, General Electric, and Johnson & Johnson as examples of companies that employ "shared value" strategies.

Porter and Kramer argue that their shared value concept goes significantly beyond corporate social responsibility, which they characterize as feel-good programs that are focused primarily on enhancing firm reputation and are not seen as integral to the firm's profitability. However, one can look alternatively at firm-wide commitment to CSR as a core company ethic, which is incentivized by more immediate marketing objectives to protect and enhance brand and reputation and which in turn drives the firm's allocation of resources and investments in ways that result in longer-term shared benefits with society.

TECHNOLOGY AND INFORMATION ASYMMETRY

High-technology markets are often characterized by information asymmetry. This is a case where one party to a transaction, such as the seller, has all the information about the underlying properties of an innovative product or service and the other party, such as the buyer, is at the mercy of the party possessing all the information. The consumer is often incapable of determining the safety of a product or service, and trusts a company to provide a safe product. The question is whether withholding relevant information from consumers that a firm is not legally required to disclose, or taking the proactive ethical route of more transparency in the information it discloses, is the better marketing strategy in the long run.

High-tech firms that have invested substantial amounts of money in research and development and seek to recoup their investment from innovative products that they are able to bring to the market first, if at all possible, lack the direct economic incentive to publicly disclose all possible problems with usage of their products if not legally required to do so. Nevertheless, these firms risk damage to their reputation if they unethically exploit their information asymmetry advantage and are exposed, a risk that has increased with the widespread publication and distribution of information on the Internet and the use of social media. We examine two examples of such behavior next.

INFORMATION ON LINKS BETWEEN CELL PHONES AND CANCER

An example of information asymmetry in the high-tech industry involves the possible relationship between cell phone use and some forms of cancer. The International Agency for Research on Cancer, a part of the World Health Organization (WHO), reported on the conclusions of the scientific literature pointing to a possible link between cell phone usage and certain forms of a malignant brain tumor.[11] However, the agency couched its warning in the acknowledgment that the evidence remains somewhat inconclusive.

The International Telecommunications Union estimated that by the end of 2010 there would be as many as 5 billion mobile subscriptions globally.[12] However, despite the increasing proliferation of cell phones, including smartphones, all over the world, it takes years for many brain tumors to develop after a known cancer-causing exposure, according to the American Cancer Society. Thus, it could be years before scientists have enough degree of certainty from a significant sampling of observable cases to merit regulatory intervention in the marketplace, other than possibly requiring warning labels.

The wireless industry's main public response to date has been to deny the existence of any cancer risks from cell phone usage. For example, John Walls, vice president of public affairs for the industry advocacy group CTIA–The Wireless Association, seized on WHO's caveats to its conclusions and dismissed the importance of WHO's report.[13] CTIA's public position is that cell phone use is entirely safe, even for children.[14] CTIA referred to the findings of the U.S. Food and Drug Administration to back up its contention.

To date, while some outside scientific studies have found possible *observational* links between cell phones and cancer, it is true that no credible publicly available human studies have appeared to prove a *causal* link between the two (Parker-Pope 2011). One ethical question facing the wireless industry, therefore, is whether it would be willing to publicly disclose the complete contents of any of its own internal studies that might show possible harms from cell phone use, or whether it will adopt the tobacco industry's model of suppressing internal reports linking smoking to cancer for as long as it could. In particular, has the industry performed any research on the potential long-term effects of cell phone use by children, which it has not shared with the public to date?

A related ethical question involving the industry's commitment to corporate social responsibility is whether it is willing to be more proactive in trying to redesign its phones to eliminate any cancer risk as much as possible, particularly to children who are using the phones. Will cell phone manufacturers and wireless service providers wait until today's generation of children using their products grow up and experience health problems that only then can finally be linked to their long-term usage, or will the companies show concern for their customers' anxieties about the possible long-term risks from cell phone usage without waiting for conclusive research that serious risks do in fact exist?

INFORMATION ON DRUG SAFETY

Another example of information asymmetry relates to the superior knowledge pharmaceutical companies possess about the safety and effectiveness of their products, as compared to that of the general public. While government regulators can intervene in the marketplace to protect consumers by requiring extensive testing before a new drug may be sold, and by requiring extensive labeling, consumers are still at an informational disadvantage vis-à-vis the pharmaceutical companies and are heavily influenced by firm branding and advertising in making their purchase decisions (Ladha 2007).

One such pharmaceutical company, Merck, jeopardized its reputation when it reportedly tried to cover up some information about the potentially harmful side effects of one of its signature products, Vioxx, a painkiller introduced into the market in 1999. Merck withdrew the $2.5-billion-a-year arthritis drug from the market in 2004, but faced thousands of lawsuits around the country alleging that the drug caused heart attacks and strokes. Its stock price took a major hit. Bloomberg Businessweek wrote at the time that "Merck's future is in question" (Bloomberg Businessweek 2004). Examples of bad publicity like the Vioxx episode may help explain the dramatic decline in public perception of the pharmaceutical industry. According to statistics cited by *Nature Biotechnology*, public approval ratings of the industry were 80 percent in 1997, but by 2004 (the year that Merck withdrew Vioxx from the market only after the reports had publicly surfaced about potential cardiovascular side effects) a majority of Americans viewed the industry as doing "a bad job" for consumers, with only 13 percent of Americans believing that the pharmaceutical industry is "generally honest and trustworthy" (Finegold and Moser 2006).

As the cellular phone and drug examples illustrate, high-tech companies face ethical decisions on when and how much to disclose publicly about their technologies and products in the absence of legal compulsion. How they decide may well impact their reputation in the market place.

CASE STUDY 1: THE ENVIRONMENT

As we mentioned at the outset of this chapter, the market fails to produce the economically efficient level of output in the face of externalities. The result may be inadequate incentives for profit-driven firms to act in a more socially responsible manner when the social costs and benefits are not factored into the prices of the firms' products and services. We have introduced the idea that some high-tech firms are responding to this externality challenge with technological and marketing strategies that visibly incorporate ethical practices.

We shall examine two case studies that illustrate these principles—one illustrating the impact of high-tech firms' behavior on the environment and the other on individual privacy. We start with the environment.

A 2011 United Nations report estimates that the incremental investment required to go "green" amounts to about 3 percent of world gross product (about $1.9 trillion in 2010) annually.[15] How can economic growth and environmental protection be reconciled as complementary and mutually supportive objectives? Is one way to decrease the adverse human impact on the environment significant investment in social innovation that takes the environment into account? What incentives do firms have to invest in such innovations that will probably not yield direct economic benefit to the firms commensurate with their investments?

SUSTAINABLE DEVELOPMENT

These questions lead us to introduce the concept of *sustainable development*, defined in the Report of the World Commission on Environment Development, a seminal

United Nations environmental document, as "development that meets the needs of the present without compromising the ability of future generations to meet their own needs . . . a process of change in which the exploitation of resources, the direction of investments, the orientation of technological development, and institutional change are all in harmony and enhance both current and future potential to meet human needs and aspirations" (United Nations 1987).

The concept of sustainable development takes economic and noneconomic factors into account, and implicates the externalities produced by technological developments and other human activities.

High-tech firms are obviously neither the primary source of global environmental harm caused by human activities nor can they be expected to completely solve the world's environmental problems. However, they do contribute to environmental harm and can also play a positive role in mitigating harms through social innovation and more environmentally conscious business practices. The reference to technological development in the Report of the World Commission on Environment and Development's description of sustainable development would necessarily encompass the activities of high-tech firms, which can have significant negative and positive externality impacts on the environment.

High-Tech Export of a Negative Environmental Externality

An example of a negative externality to the environment created by high-tech firms involves the disposal of firms' electronic products at the end of their product life cycles. The United States alone disposes of about 100 million television sets and computers each year, according to the international Commission for Environmental Cooperation (Ramstack 2011). Products with cathode-ray tubes, such as televisions and computer monitors, can be especially harmful to humans and the environment, according to a 2008 report prepared by the U.S. Government Accountability Office (GAO) for the House of Representatives (U.S. General Accounting Office 2008).

Some high-tech firms have used what turned out to be unethical scrap recyclers to dispose of their products or product parts. The recyclers ship the electronic products or parts to undeveloped countries, where they are handled and disposed of in a manner that creates environmental and health hazards from lead, cadmium, mercury, and other so-called e-wastes that can contain cancer-causing toxins.

Puckett (2006), who belongs to the watchdog group Basel Action Network, wrote about how U.S. electronics firms have turned a blind eye to unethical recyclers, effectively allowing the export of their end-of-life products or product parts to some developing countries willing to receive them for disposal, irrespective of the environmental and health hazards from toxic wastes. The Basel Action Network is named after the Basel Convention, which established an international legal regime governing the export and import of hazardous wastes for disposal, stipulating that a country may ship hazardous waste only after receiving prior written consent from the receiving country. The United States has signed but not actually ratified the Basel Convention treaty. Neither the treaty nor U.S. hazardous waste regulations have deterred exports of potentially hazardous used electronics.

Firms may be technically complying with the law, but their indifference to where their electronic products end up means that they are externalizing their end-of-life product costs onto the backs of these recipient countries' citizens by exposing them to e-wastes. As long as electronics firms can legally hand off their end-of-life products and product parts to unethical recyclers who operate under the radar in disposing of them to developing countries willing to receive them, what private economic incentives do these firms have to change the status quo?

In the absence of applicable legal requirements, for example, what incentives are there for high-tech firms to develop upstream innovative solutions through greener product designs, toxic elimination, and innovative manufacturing and life-cycle processes that would enable products such as computers to be made, used, and returned to the producer in a closed-loop system (Puckett 2006)?

One incentive, perhaps, is to avoid the reputational harm electronics firms risk to themselves and to their industry as a whole, to the extent they become associated in the public's mind with the behavior of the unethical recyclers that watchdog groups such as the Basel Action Network are monitoring and widely publicizing. Puckett (2006), who has highlighted the unethical behavior of some recyclers on television shows such as *60 Minutes*, has also criticized what he called the electronic industry's "tacit stratagem . . . to avoid both accountability and real downstream cost for its hazardous materials use and poor end-of-life design considerations."

More attention from the mass media and the global platform of the Internet, including blogs and social media, can lead to widely publicized embarrassing revelations, resulting in potential harm to the electronics industry's reputation.

There is also the carrot of favorable publicity for socially responsible action. To this end, the Basel Action Network has launched a program to use third-party auditors to certify recyclers who don't export hazardous electronic waste. It is trying to obtain assurances from electronics companies to use certified recyclers whenever possible. Its e-Stewards Initiative includes companies that have committed to environmentally responsible disposal of e-wastes. For example, the electronics firm Samsung's logo is displayed on the e-Stewards Initiative Web site.[16]

HIGH-TECH ENVIROPRENEURIAL MARKETING

Continuing with the electronics product recycling example, some companies have sought to develop proactive marketing strategies around environmental protection that demonstrate their commitment to corporate social responsibility. Sony, for example, decided to launch its own nationwide electronics recycling take-back program across the United States, where customers can drop off any Sony product for free. A senior executive at Sony explained to *ComputerWorld* that his company's recycling take-back program is not only good for the environment but also creates "brand loyalty" (Gross 2008).

Varadarajan (1992) has used the term *enviropreneurial marketing* to encompass such "environmentally-friendly marketing practices, strategies, and tactics" that are designed to do well by doing good—achieving competitive differentiation advantage for the firm's products while reflecting the importance the firm attaches to its corporate social responsibility. Menon and Menon (1997) took this concept one step further in tying enviropreneurial marketing to the corporate social performance model, where companies integrate their business objectives with their social performance goals.

Baker and Sinkula (2005) concluded from their empirical studies that (1) a firm's enviropreneurial marketing strategy development is "positively related to new product success," and (2) new product success is "positively related to change in market share" (470–471). They view enviropreneurial marketing as "a unique organizational resource, one that is difficult to learn and costly to copy" (471). The greater the degree of such tacitness of a resource (tacitness is discussed in chapter 1), the greater its difficulty to learn and cost to copy. Along with other firm resources and capabilities, such as strength of distribution and pricing power, enviropreneurial marketing helps contribute to (but does not in itself directly drive) a firm's change in market share (Baker and Sinkula 2005).

New technology resulting from social innovations may also be a source of enviropreneurial marketing in an emerging market for innovative green solutions, such as the alternative energy market discussed later in this chapter. Firms willing to engage in the risk of social innovation and enviropreneurial marketing may capitalize on their first-mover advantage if they are successful (Cronin et al. 2010).

ENERGY, CARBON GASES, AND CLIMATE CHANGE

"Many energy production and consumption activities result in negative externalities, perhaps the most recognized being environmental damage," reads a 2010 Congressional Research Report (Sherlock and Marples 2010, p. 2). The production and consumption of energy primarily uses fossil fuels, resulting in associated greenhouse gas emissions that have contributed to potentially the most disastrous negative externality of all to the environment—global climate change.

Worldwide demand for energy is increasing at a rapid pace, driven most recently in part by significant growth of usage in big emerging-market economies such as China and India. Akpan's analysis found that since approximately 1850, "global use of fossil fuels (coal, oil, and gas) has increased and dominated world energy consumption and supply" (Akpan 2012, p. 21). The consequence is a rapid growth in the emission of carbon dioxide and other carbon-based greenhouse gases into the atmosphere. Elevated concentrations beyond what natural phenomena would produce are altering the earth's climate in ways that can have harmful consequences.

Since climate change is global in scope, the United Nations and other transnational forums have spearheaded efforts to study greenhouse gas emissions and to devise programs to reduce them. These efforts accelerated after the 1992 UN Conference on Environment and Development concluded with the signing of the Rio Declaration, Agenda 21, and the Framework Convention on Climate Change (UNFCCC). Agenda 21 promotes the development, transfer, and use of improved energy-efficient technologies, the application of economic and regulatory means to mitigate environmental and other social costs, and other energy efficiency–related measures. The UNFCCC attempted to set enforceable targets for emission reductions for the post-2000 period, culminating in December 1997 with the setting of targets for emission reductions for the years 2008–2012. The resulting treaty is known as the Kyoto Protocol, but neither the United States nor major energy-consuming nations in the developing world such as China and India have ratified the treaty. Attempts to negotiate binding emission reduction targets for the post-2012 period, including the United States, China, India, and other nonparticipants in the Kyoto Protocol, have run into political obstacles.

Meanwhile, the Intergovernmental Panel on Climate Change (IPCC) was organized as the leading international body for the assessment of climate change. It was established by the United Nations Environment Programme and the World Meteorological Organization to provide the world with a clear scientific view on the current state of knowledge in climate change and its potential environmental and socioeconomic impacts. It taps the expertise of thousands of climate change scientists worldwide. By serving as the focal point to organize the studies, report on the results, and reflect broad consensus among the world's leading climate scientists, the IPCC hopes to influence policymakers to take action and to embed into the public consciousness the critical importance of furthering the social norm of responsible sustainable development.

After extensive scientific study, the vast consensus in the scientific community, as reported by the IPCC, is that most of the observed increase in global average temperature since the mid-twentieth century is very likely due to the "observed increase in anthropogenic greenhouse gas concentrations" (IPCC 2007). The IPCC concluded that these human-produced concentrations are accumulating to dangerous levels that may cause irreversible climate changes harmful to all living species and to the planet Earth itself. The findings of the IPCC are summarized in Exhibit 7.1.

HIGH-TECH INDUSTRY IMPACTS ON GLOBAL WARMING

High-tech manufacturing processes and operations that rely principally on fossil fuels as their energy source contribute to fossil fuel–induced carbon dioxide emissions, which in turn contribute to global climate change. For example, according to SMART 2020, a detailed study conducted by the Climate Group, the world's electronic gear (including PCs, their peripherals, telecoms networks and devices, and data centers) produced about 2 percent of total emissions from human activity in 2007 (The Climate Group 2008).

The Climate Group is a global independent, not-for-profit organization that brings together business and government leaders to address the problem of climate change and devise possible solutions.[17] The group undertook its study and prepared its report on behalf of the Global eSustainability Initiative, an international strategic partnership of information and communications technology companies (ICT) such as Cisco, Ericsson, Hewlett-Packard, Intel, Microsoft, Motorola, Research In Motion, Alcatel-Lucent, Nokia, Verizon, AT&T, Sprint, and Vodaphone, as well as industry associations that have set out to foster sustainable development through global cooperation and promoting carbon-reducing technologies.

The study quantified the direct emissions from ICT products and services based on expected growth in the sector. It also examined where information and communications technology companies could enable significant reductions of emissions in other sectors of the economy such as smart motor systems, smart logistics, smart

Exhibit 7.1 **Intergovernmental Panel on Climate Change**

"Global atmospheric concentrations of carbon dioxide, methane and nitrous oxide have increased markedly as a result of human activities since 1750 and now far exceed pre-industrial valued determined from ice cores spanning many thousands of years."

"The primary source of the increased atmospheric concentration of carbon dioxide since the pre-industrial period results from fossil fuel use, with land use change providing another significant but smaller contribution."

"Most of the observed increase in globally averaged temperatures since the mid-twentieth century is very likely due to the observed increase in anthropogenic greenhouse gas concentrations."

Source: Intergovernmental Panel on Climate Change (2007).

buildings and smart grids, and quantified these in terms of CO_2 emission savings and cost savings.

The SMART 2020 study estimated there would be nearly a tripling of the information and communications technology industry's carbon footprint between 2002 and 2020. The study found that one-quarter of the carbon emissions generated came directly from the materials and the manufacturing themselves. Three-quarters came from usage of the ICT products and services.

The SMART 2020 study estimated the number of personal computers globally is expected to increase from 592 million in 2002 to more than 4 billion in 2020. In order to reduce the total carbon emissions of personal computers predicted for 2020 to below 2002 levels, there would need to be a 95 percent efficiency improvement in the overall impact from PCs, which the Smart 2020 study says would require transformational technology, not just increased energy efficiency and longer product life.

Add to the personal computers the increasing use of data centers that store data and videos for user retrieval and house clusters of servers, storage devices, networking equipment, power supply, and cooling equipment. According to the SMART 2020 study, if growth continues in line with demand, the world will be using nearly seven times the number of servers in 2020 as it did in 2007. Data centers are the fastest growing contributor to the information and communications technology industry's carbon footprint.

However, the SMART 2020 study also pointed to potentially positive externalities created by the information and communication technology industry's products and services. The study estimated that by 2020 the ICT industry could enable the reduction in the amount of emissions from other industries by five times the carbon footprint of the ICT firms themselves. Even without considering the role that alternative energy technologies could play in future manufacturing processes, ICT firms' products and services are already helping to reduce emissions in other industries through such means as enabling videoconferencing to replace the travel required for face-to-face meetings. Other future "enabling effects," as the SMART 2020 study calls them, could include a "smart" electrical grid's application of data networking technology to more efficiently manage demand and energy consumption.

The SMART 2020 study also mentioned promising breakthrough technologies that hold the potential of reducing the carbon footprint of the ICT products themselves, such as "solid state hard drives, which could reduce energy consumption by up to 50 percent, choleristic LCD screens that reduce monitor energy consumption by up to 80 percent and direct methanol fuel cells that can deliver 20 percent savings for power supplies" (The Climate Group 2008, p. 21). Quantum and optical computing may also produce positive impacts, according to the study.

The Global eSustainability Initiative's ICT members have committed to implement the specific recommendations of the SMART 2020 study, which include "reducing the carbon footprint of the ICT sector and realising the opportunities identified to enable emissions reductions in other sectors."[18]

ALTERNATIVE ENERGY INNOVATION

New research and development into alternative sources of renewable energy that do not create carbon emissions would be more incentivized in the present market economy if firms and individuals would have to take the cost of carbon emissions into account when using fossil nonrenewable fuels—that is, if the negative externality of carbon emission were included in the final price of the product or service contributing to the emission. Without a predictable floor for prices of high-carbon energy sources, and no practical means in place for internalizing the negative externality, there may not be enough incentive for private firms alone to engage in expensive and risky R&D projects to develop alternative energy technologies.

Current public policy attention has focused on measures to bring about the internalizing of the negative externality, such as market-based "cap-and-trade" strategies or carbon taxes to increase the direct costs to the emitters of greenhouse gases. They are intended to provide a market-based price signal to private firms and individuals about the external cost of their activities in relation to the environment.

In a cap-and-trade program, the regulator sets an overall cap on emissions and grants permits for allowable emissions. Individual firms then trade the permits among themselves, transferring permits from firms that do not need their full allotment to those who need more than their original allotment of emissions allows (Leggett 2009, p. 21). Under a system using carbon taxes charged to a source of emissions according to its total emissions, the objective is to more directly impose the external costs of greenhouse gas emissions on the source of emissions, although it is possible that some or all of the tax will be passed on to consumers.

The idea behind both proposals is to encourage producers and users of energy to move toward alternative renewable sources of energy by making them more economically viable in comparison with the "re-priced" carbon-based sources of energy. However, such proposals have faced political obstacles to being implemented, particularly in the most significant energy consuming countries, the United States and China.

Tax incentives are granted in the United States to individuals and businesses to make investments in energy efficiency or renewable energy generation that meet certain criteria. The Congressional Research Service has also reported that "Federal appropriations of billions of dollars have been enacted in recent years to stimulate more efficient energy technologies. . . . These subsidies can take the form of tax credits for R&D, cost-sharing grants or contracts, direct investments, loan guarantees and others" (Leggett 2009, p. 24).

However, these tax incentives and other forms of government assistance to support R&D in alternative energy technologies, such as solar and wind energy, have not reached a tipping point sufficient to spur major private firm investments by U.S. companies in these technologies. China, on the other hand, has been very proactive in fostering such R&D (Campbell 2010).

In the absence of sufficient government incentives to develop alternative technologies, and in the absence of a compelling business case for high-tech firms to invest significant amounts of money of their own to develop technology that may not prove to be commercially viable, little progress can be expected in the private sector for the foreseeable future unless some vanguard leaders emerge to invest in what we have referred to as social innovation.

A firm may be more inclined to take that leap if it sees a marketing advantage in being identified with a rising public consciousness that views global climate change as a real health and ecological threat that must be dealt with on an urgent basis. This public consciousness, in turn, requires the prevalence of social norms that feature "environmentally responsible decisions as more socially acceptable" and where "less responsible decisions are stigmatized" (Blodgett and Parker 2008, p. 31). Wiser and Pickle (1997) advise that "marketing messages and product positioning should be targeted to the most effective forms of social pressure and social norms" (pp. 20–21). They also cite studies by the Roper Organization and others that identified different market segments (i.e., groups of customers with similar needs) that should be targeted in the marketing of alternative "green" energy products according to types of customer values and primary motivations. This is an example of product positioning discussed in the chapter on product strategy (i.e., the way a firm communicates the relative benefits of its products to customers, which may vary across different market segments).

Environmentally conscious customers might be targeted by messages that emphasize personal responsibility for the state of the environment and its impact on the well-being of future generations, which may include their own children and grandchildren.

Customers who are most influenced by word of mouth and peer pressure (see chapter on product strategy) might be more effectively targeted by messages that emphasize who else is moving toward the use of green energy sources—what Wiser and Pickle (1997) describe as targeting "through interpersonal influence via peer pressure" (p. 21). They conclude from their review of behavioral studies that for such customers "marketing messages should emphasize that others are participating or plan to participate, therefore exploiting the fear of social sanction to establish a 'bandwagon' effect" (p. 21).

As we discussed in chapter 1, bandwagon effects, also known as network externalities, arise when the value of a product or service to an individual user rises as the number of users increases. As we also noted, in order to succeed, bandwagon products and services need to reach a critical mass, without which start-up problems arise. The challenge for high-tech firms is how best to help build demand for their alternative energy products even before a critical mass is reached. To what extent do they want to take the market risk of going out in front with investments in R&D and marketing before alternative energy technology is widely diffused, in the hopes that they will reap the potential rewards of enhanced reputation and positive product brand identification with alternative energy if and when the critical mass is reached?

An example of a firm that was evidently willing to take such a risk is the Danish wind power company Vestas, whose slogan is "Wind. It means the world to us." Vestas has what it describes as "30 years of pioneering expertise" in wind as an energy source. Its Web site declares its goal: "Being the market leader within wind power, Vestas aims to create the world's strongest energy brand."[19]

Vestas employs more than 20,000 people globally and has production facilities in more than a dozen countries, including the United States and China. While it has grown to be the world's largest manufacturer, seller, installer, and servicer of wind turbines, its market share has declined in recent years as competitors such as General Electric have entered the market and some price wars have ensued. In response, Vestas's management has decided to place less emphasis on market-share gains through aggressive price discounts and concentrate on "looking after the customers" and making further technological breakthroughs (Reddall 2010). Thus, after its early leadership in the alternative energy wind market as a result of its "pioneering" activities, Vestas would appear to be relying on market-sensing and customer-linking capabilities (discussed in chapter 1) to sustain its competitive advantage.

CREATING SUSTAINABLE DEVELOPMENT SUPPLY CHAIN NETWORKS

Some leading purchasers of high-tech products and components, including some high-tech firms themselves who depend on their suppliers in producing their products, are using their purchasing power with their suppliers to extend the reach of their own codes of best practices regarding the environment. The purchasers inform potential suppliers that, as customers, they favor vendors who provide environmentally sustainable products, systems, technology, and services, who are environmentally sensitive in their daily operations and who have established environmental management programs. This begins the process of establishing green supply chain partnerships, which the supply chain literature has studied under the rubric of "supply construction" (Cronin et al. 2010).

Customers with significant purchasing leverage are increasingly incorporating a code of environmental best practices into their purchasing contracts, which may be based on standards promulgated by recognized organizations such as the International Organization of Standardization (discussed below). In the contract, a supplier may be asked to commit, during the term of the purchase contract, to work toward a measurable reduction in its energy and resource consumption and to control significant emissions and discharges at the supplier's own facilities, all in accordance with the customer-specified best practices. These provisions may be backed up by performance metrics and monitoring mechanisms, including the customer's right to conduct its own periodic on-site audits to ensure compliance with the customer-specified environmental best practices, at least as they relate to the products and/or services delivered to the customer.

Finally, the supplier may also be required by its purchase contract with the customer to enter into contracts with its subcontractors and other vendors in its own supply chain that demand the same environmental best practices requirements.

The contracting mechanism of committing the purchaser's supply chain, and the supply chains of the purchasers' suppliers, to environmental best practices has the bandwagon effect of building a network of similarly minded interdependent companies along an entire vertical value chain. The network can also extend horizontally, linking high-tech industry peers in organizations focused on sustainable development, such as the information and communication technology industry's Global eSustainability Initiative discussed earlier. Through participation in such organizations, high-tech firms are able to move toward standardization of environmental best practices for themselves and their suppliers, using common supplier questionnaires, risk assessment, auditing, online training and data management tools for communicating with and monitoring their suppliers.[20]

In another move toward standardization of environmental best practices, the International Organization of Standardization (ISO) has developed Standard 14001 for an environmental management system, which requires complying organizations to determine their impact on the environment, create a plan to minimize the environmental impact, and monitor the effectiveness of the system at meeting objectives. While organizations can self-certify their conformance to the standard, some firms choose to have their systems "certified by third-parties to add further validity to their environmental efforts" (Cronin et al. 2010, 165). Moreover, if a major customer requires an ISO 14001 certificate of conformance with the standard, third-party certification will likely be mandatory.[21]

Global networks enabling the sharing of information on supply chain best practices are also sprouting up. For example, a Sustainable Supply Chains Web site was launched at the 2010 UN Global Compact Leaders Summit as a central resource to simplify the research task of businesses seeking information about supply chain sustainability. The site also invites organizations, companies, and others to share information about their supply chain initiatives and resources. See Exhibit 7.2 for the UN Global Compact's statement of the business case for incorporating environmental and other CSR considerations into supply chain management.

The Global Reporting Initiative (GRI) is a network-based organization that, according to its Web site, "pioneered the world's most widely used sustainability reporting framework." GRI has formulated principles and standardized indicators for organizations to use in measuring and reporting their economic, environmental, and social performance. This program of standardized reporting includes a focus on disclosure of performance with respect to supply chains.[22]

In sum, we have offered in this section illustrations of two interrelated trends in the area of supply chain management and the environment that can affect high-tech firms' corporate social responsibility marketing strategies: (1) the bandwagon effects

Exhibit 7.2 **United Nations Global Compact: The Business Case for Supply Chain Sustainability**

"Incorporating environmental, social and governance considerations into supply chain management can deliver a range of business benefits:

- Risks are better anticipated and managed (risk is spread out across different players)
- Reduced operational risks such as disruption to supply, increased cost and lack of access to key raw materials
- 'Informal' or 'social' license to operate within communities, legal systems and governments that otherwise might be antagonistic
- Reduced costs and enhanced efficiency and productivity
- Improved working conditions can reduce turnover and improve quality and reliability
- Environmental responsibility improves efficiency and profitability
- Corporate brand and values, and customer and consumer confidence and loyalty are protected and enhanced
- Process and product innovation. Empowered suppliers uncover opportunities for developing sustainable products and services
- Examples from leading companies show that good supply chain management can increase shareholder value"

Source: United Nations Global Compact Web site: http://www.unglobalcompact.org/Issues/supply_chain/index.html.

created by expanding sustainable development supply chain networks, and (2) increased standardization of environmental best practices, which together with publicly available reporting and certification of company performance, provide customers with valuable information on the firm's commitment to CSR.

CASE STUDY 2: PRIVACY

The ease with which government and private sector entities can obtain personal information about individuals on the Internet, and possibly use that information to the individuals' detriment without their knowledge, much less consent, raises a host of legal and ethical concerns about privacy.

As we discussed in chapter 4, laws and regulations in the United States and other countries provide varying degrees of protection for personal identifying information. Businesses that rely heavily on electronically stored personal identifying information, such as name, address, and account numbers, for marketing purposes must be particularly mindful of how privacy laws and regulations in different countries can impact their marketing strategies. High-tech service providers that help businesses process and analyze their stored data are also directly impacted by the legal and regulatory privacy landscape.

WEB SITE USER TRACKING

However, there is another area of Internet commerce where the law lags behind the development of technologies that have privacy implications—systems that can track users'

Web site visits. Since the owners of many Web sites on the Internet still provide free access to users, they rely on advertisements placed on their Web sites to produce revenue. Advertisers will pay a premium if they know that the advertisements they are purchasing will be seen by a target audience most likely to be interested in the product or service offered. This is where sophisticated Web site tracking technologies come into play.

Yet while the developers of the tracking software, as well as the firms that compile and report the results of the tracking and the advertisers and Web sites that use those results to target their advertising all reap commercial benefits, there is a negative externality that is imposed on unsuspecting users. How should firms handle the ethical and marketing implications of Web site tracking technology?

THE TECHNOLOGY

Technology has been developed to enable the targeting of online advertisements directly at individual users based on their Web surfing activity. This practice is widely known as "behavioral" or "e-havioral" advertising (Ruane 2009).

"Cookies" (the small text files downloaded from a Web site visited by a user to the user's computer that can store retrievable information) are a relatively common means of keeping track of information on user visits to Web sites.

According to a July 2010 investigative report in the *Wall Street Journal*, the top fifty Web sites in the United States install an average of 64 individual trackers to visitor computers with little to no disclosure (Angwin 2010). Researchers at AT&T Labs and Worcester Polytechnic Institute found tracking technology on 80 percent of one thousand popular sites (Angwin 2010).

There are companies that specialize in using sophisticated software to capture users' browsing patterns and even their keystrokes while on the Internet. Such a company "packages the data into profiles about individuals, without determining a person's name, and sells the profiles to companies seeking customers" (Angwin 2010).

The tracking software uses a unique alphanumeric code associated with the user's device for user identification purposes, purportedly without revealing any personally identifiable information such as the user's name, address, social security number, or bank account numbers. Marketers who value what this technology has to offer are not looking for the user's personal identity information. They want to be able to match the user's interests and preferences, as evidenced by the user's Internet tracks, with targeted ads that can then be displayed on, or sent electronically to, the user's computer.

Sometimes Web sites distribute the tracking software to users' computers for a fee paid by the company that developed, owns, or licenses the software. However, sometimes, according to the *Wall Street Journal* report, not even the Web site owners know that they are being used as distribution vehicles for tracking company software placed on the users' computers (Angwin 2010).

In addition to using cookies, tracking technology is getting even more intrusive. The Angwin's (2010) *Wall Street Journal* investigation found "new tools that scan in real time what people are doing on a Web page, then instantly assess location, income, shopping interests and even medical conditions."

The tracking technology enables advertisers to follow people around the Internet. Entering into partnerships with users' Internet service providers, advertisers can gain access to all Web activity in which an Internet service provider's subscribers engage. The information collected is real-time and detailed, and it is marketed on the Internet. User data is collected in multiple contexts and at various times throughout the day, such as when users browse Web sites, purchase items online, or use a geolocation application on a mobile device. Companies that collect data through such activities share the data with multiple entities willing to pay for such market intelligence, including third parties far removed from direct business interaction with the users themselves.

Although the European Union has tried to address the proliferation of cookies with a new requirement that a user provide express "opt-in" consent before a Web site can place cookies on the user's computer, technology is at least one step ahead of the law. Moreover, there is no such legal requirement as of yet in the United States, although the Federal Trade Commission (FTC) has recommended guidelines for businesses to consider implementing in connection with their collection and use of consumer data from Internet tracking technologies (Federal Trade Commission 2010).

ETHICAL ISSUES

Not only the law, but industry ethical practices too are not able to fully keep up with the impact on consumer privacy caused by the proliferation of tracking technologies. Indeed, this may be an area where a firm's self-imposed ethical limitations on the tracking technologies it uses could deprive that firm of valuable marketing intelligence for a competitive edge. The software companies that develop the tracking systems would argue that they are simply filling a need of marketers who want the best "bang for the buck" for their investment in online advertising. The marketers, and the advertising industry that supports them, would argue they are helping consumers by enabling them to receive information via the Internet that they are likely to be interested in, rather than a flood of useless, unfiltered information. But at what negative externality cost to consumers' privacy?

The collection and commercial use of data about their interests and preferences, gleaned from tracking of their Web surfing, is often invisible to consumers using the Internet and beyond their control. The recipients of the data possess an intimate picture of an individual's life, which they are free to exploit for commercial purposes.

Assume that effective technology was in place to prevent any sensitive personally identifying information of Web users from being accessible to the advertisers and that no laws were being broken by tracking the Web-surfing habits of an individual Internet user, identified only by alphanumeric codes, for the purpose of creating a profile of that individual's interests and preferences for targeted marketing purposes. What are the ethical issues in using technology to track, gather, compile, and sell information about the individual's surfing habits, without the individual's knowledge, to enable the delivery of information that may be of interest to that person?

On the one hand, it can be argued that once you decide to use the public Internet you give up a measure of privacy regarding which sites you are visiting. You are getting information for free from a Web site in return for enabling the site's primary source of revenues—its advertisers—to know the computer address of those who visit the site. Moreover, innovations that consumers value, such as mobile and social networking services, are funded, in part, by the growth of targeted advertising that relies on commercial use of consumer data.

There may not be as much of an ethical issue if we are dealing with so-called "first-party" trackers. They gather individuals' usage data regarding their sessions on the Web site where the individuals being tracked have initiated the sessions themselves. If those visitors return to the site, their user experience will be improved because of the first-party trackers, as they navigate the site more quickly and can take advantage of its features such as auto-complete forms and shopping carts for purchases. As long as the Web site discloses that information regarding the user's interaction is being collected by the site for subsequent commercial use, the user has a choice as to whether he or she is willing to share certain information in exchange for the services offered by the site.

On the other hand, so-called "third-party trackers" are more likely to be perceived by regulators and consumers as pushing the envelope when it comes to user privacy and online ethics. These trackers include advertising networks and data companies (including ad exchanges and data aggregators) that collect Web-viewing data across multiple unaffiliated sites and use such data to serve the purposes of online targeted advertising. They use the data to create user profiles, predict user preferences, and sell the results to advertisers who want to show users ads that are most likely to be of interest to them.

There is a sense that, with technology enabling all our Web searches to be continuously tracked by such third-party trackers as we travel from site to site, "Big Brother" is watching our every move on the Internet whether we know it or not. The focus is not limited to gathering information on how an individual uses a particular site to improve the user's direct interaction with that site. Instead, the focus is on building a rolling profile of the individual based on everything the individual does while on the Internet. A majority of Internet users do not know they are being tracked so extensively in real time, nor do they have any idea where the detailed dossier put together from the tracking about their interests, preferences, and the like ends up.

MARKETING INCENTIVES TO CURB WEB SITE TRACKING

High-tech companies that produce the tracking software, and the compilers and brokers of user profile information gathered from the tracking, have little incentive to curb their tracking-enabling activities on their own as long as the activities remain legal and they have a market of advertisers interested in the results they are able to track and compile. This is particularly the case with non-consumer-facing software developers and information aggregators, of whom consumers may know little or nothing.

However, prominent consumer-facing Web sites themselves, as well as the marketers who serve them, and the leading browser vendors and search engines with whom consumers regularly interact, have begun to feel some public backlash as the privacy implications of their role in tracking become more well known. In an effort to dispel privacy concerns of some Internet users and to avoid new regulations by the Federal

Trade Commission or other government agencies on the EU model, media and marketing associations are trying to get out in front of the issue and self-regulate, using, for example, guidelines recommended by the FTC (Federal Trade Commission 2010). They are promoting a voluntary program that offers "consumers enhanced control over collection and use of Web viewing data for online behavioral advertising" (American Association of Advertising Agencies 2010). Their stated rationale is that providing consumers with choices about online advertising is essential to building the trust necessary for the online consumer market to grow, which requires an industry-wide effort to be effective.

In addition to advertising agencies and media companies, participants in this program include prominent search engine companies such as Google and Yahoo, telecommunications companies such as AT&T and Verizon, and technology companies such as Microsoft, Dell, and Research in Motion (RIM), along with automotive companies, financial services companies, and consumer goods companies.[23]

The associations have recommended the use of an "Advertising Option Icon" and accompanying language, to be displayed "within or near online advertisements or on Web pages where data is collected and used for behavioral advertising." Users can click on the icon and be taken to a full disclosure page and an opt-out option (American Association of Advertising Agencies 2010). Their effectiveness, however, depends on how clear the disclosures are, how easy the opt-out option is to use, and how broadly such mechanisms are implemented across the major Web sites used by consumers and the marketing entities that support them (FTC 2010).

In addition, technology itself may offer some solutions to the privacy concerns that tracking technologies have created. For example, major browser vendors are offering technical mechanisms to limit online tracking, with varying scope and ease of use. One variation is a browser extension that means an Internet user can easily opt out of personalized advertising from all participating ad networks only once and store that setting permanently. This is an example in which some leading technology companies that are consumer facing, such as Google and Microsoft, see a marketing benefit in offering and promoting a technological solution that gives consumers more control over the negative effects they experience from online behavioral advertising (Google Public Policy Blog 2011).

In sum, online tracking technology has created a negative externality in the form of a threat to Internet user privacy that the law alone has been incapable of addressing. The tracking technology developers, information compilers, and brokers, who are largely invisible to consumers, have done little on their own to address the issue. However, in response to the threat of consumer and regulatory backlash, prominent Web sites, advertisers, Web browser and search engine firms who have more direct interaction with consumers have taken voluntary steps to address the effects of this negative externality. The decision of such firms to forego certain commercial benefits to themselves from customer tracking information in favor of minimizing tracking technology's social costs to individual privacy is not an altruistic one. This ethical choice is driven in part by marketing strategists' concerns with negative impacts on firm-wide reputation and branding.

BUSINESS ETHICS BEST PRACTICES CHECKLIST

The purpose of this chapter has been to demonstrate how negative and positive externalities created by high-tech firms relate to the strength or weakness of their business ethics practices, which in turn can affect their marketing strategies by protecting their reputations and the long-term quality of their brands.

We have seen illustrations of high-tech companies' ethical choices in dealing with such issues as sales tactics, product safety disclosures, environmental impact, tracking of Internet users' online behavior, supply chain management, and overall commitment to corporate social responsibility. It is useful to close this chapter by outlining a checklist or framework for a firm to consider in devising and implementing a comprehensive top-to-bottom ethical program.

Such a program should have at least two major components. The first focuses on the honesty of the firm in the conduct of its business and how the firm codifies and implements written standards toward that end. As we previously mentioned, these are baseline business ethical practices that cut across all industries. The second ethical component focuses on the firm's level of corporate social responsibility, which we have concentrated on in this chapter because of its particular relevance to how high-tech firms can deal with externalities and turn them to their marketing advantage. The proposed ethical practices checklist is as follows:

1. *Firm should have a written code of firm-wide ethical standards for honest behavior and an organizational structure committed to by the firm's top leadership for its successful implementation.* Does the firm have a written code of ethical standards of conduct that covers all aspects of honest dealings in its day-to-day business, including marketing practices, corporate governance, conflicts of interest, and the behavior of its managers and employees toward each other and toward shareholders, customers and potential customers, suppliers, and regulators? Has the firm assigned high-level personnel to oversee compliance with such ethical standards? Is the top of the firm's management leading by example?
2. The mechanisms the firm has in place to implement its ethical code and compliance should at minimum include:

    - Frequent communication of the ethical standards and their centrality to the firm's core mission and values firm-wide, including employee training on expectations for acceptable behavior;
    - Regular audits for compliance with the standards;
    - Enforcement through appropriate disciplinary actions and compensation structure that does not reward bad behaviors such as dishonest sales tactics;
    - Honest and accountable corporate governance, including systems and procedures to ensure accurate, complete, and transparent accounting and public disclosures of information about the firm's products, business, and finances;

- Healthy and honest work environment in which employees are treated respectfully and fairly;
- Easily accessible ethics phone hotline to ask questions and report infractions; and
- Ethical behavior screening and appraisal criteria for employees and suppliers.

3. *A firm should build and maintain a firm-wide culture of corporate social responsibility*. Does the firm consider the societal impacts of its research and development, production processes, products, and marketing/sales practices during every stage of R&D, production, marketing, and product life cycle? To what extent has the firm engaged with stakeholders such as NGOs and residents of the locations where the firm does business to obtain their input in taking practical steps to minimize social harm and maximize social value from its activities? What capital and human resources is the firm willing to devote to social innovation and enviropreneurial marketing?

We close this chapter with the following thought:

If ethics are poor at the top, that behavior is copied down through the organization.
—Robert Noyce, inventor of the silicon chip

SUMMARY

In this chapter, we have examined how the use of technology by firms and entire industries in their businesses can produce social costs such as environmental harms, and social benefits such as improvements in delivery of health care to impoverished regions of the world. We have also discussed how high-tech firms are dealing with these social costs and benefits, known as externalities. We considered how high-tech firms' ethical choices—the selection of business practices that are consistent with a firm's normative standards of acceptable behavior—address such externalities beyond what the law may dictate.

We observed the impacts of a firm's ethical choices on its reputation and brand. We discussed and provided examples of how a firm's public demonstration of ethical behavior plays into its marketing strategies to enhance company reputation and improve customer image of its brand, behavior that would in turn help mitigate negative externalities and internalize positive externalities even in the absence of effective legal rules and enforcement mechanisms.

We noted that even if firms do not necessarily expect to be able to quickly recoup the direct costs of developing and offering their customers "green" and other socially responsible technologies, the marketing and competitive benefits that flow from being able to demonstrate "best in class" corporate ethics to their stakeholders, such as customers and employees, provide incentives for socially responsible behavior that help to overcome the market failure posed by externalities. The business benefits we discussed include "brand insurance" and protection of reputation, employee recruitment and retention, and fostering competitive growth in emerging markets.

We used two case studies—the impact of high-tech firms' behavior on the environment and individual privacy—to demonstrate how some high-tech firms are responding to the externality challenge with technological and marketing strategies that visibly incorporate ethical practices.

KEY TERMS

brand insurance
corporate social responsibility
"enviropreneurial" marketing
externality
information asymmetry
market failure
reputation insurance
shared value
social alliances
social goods
social motivation
sustainable development

DISCUSSION QUESTIONS

1. Why are externalities associated with the economic concept of *market failure*?
2. How can the use of technology create negative externalities, and what are some examples?
3. What is the relationship between positive externalities and social innovation? Describe an example of a social innovation in the environmental field.
4. Define *corporate social responsibility* (CSR).
5. Describe and provide examples of how CSR provides high-tech firms with "brand insurance" and "reputation insurance."
6. What is *enviropreneurial marketing* and how does it relate to the concepts of tacitness and first-mover advantage?
7. Describe the high-tech industry's impact (negative and positive) on climate change.
8. How should marketers of alternative energy products target different market segments, and how do "word of mouth" and peer pressure affect their marketing strategy?
9. How do sustainable development supply chain networks develop and create a bandwagon effect?
10. Suppose that a smart digital watch is developed with wireless capabilities, built-in sensors, and connection to digital networks that can continuously record an audio and visual record of the wearer's activities streamed to social networks and archived for later retrieval. However, other people with whom the wearer of the smart digital watch interacts may not be aware that they, too, are being recorded in real time. What ethical and reputational issues would social network providers face in allowing streaming that contains audio and images of people recorded without their knowledge or consent, and what strategies would you advise these providers to follow in response?

NOTES

1. See Gearing up: From corporate responsibility to good governance and scalable solutions, United Nations Global Compact on Sustainability (2004). Available at http://www.unglobalcompact.org/docs/news_events/8.1/gearing-up.pdf.

2. The findings reported here and displayed in Figures 7.3 and 7.4 are reproduced with permission of GlobeScan Incorporated: London, Toronto, San Francisco.

3. See Corporate Branding Social Responsibility Survey 2010. Survey conducted by Penn Schoen Berland in conjunction with Burson-Marsteller and Landor at http://www.slideshare.net/BMGlobalNews/csr-branding-survey-2010-final.

4. See Why CSR matters to corporate reputation, reputation intelligence, from the Reputation Institute (Spring 2010). Available at http://www.reputationinstitute.com/knowledge-center/intelligence_files/Rep_Intel_Why_CSR_Matters_in_2010.pdf.

5. 2012—The Global RepTrak™ 100: The World's Most Reputable Companies. Reputation Institute, p. 22. Available at http://www.rankingthebrands.com/PDF/2012%20RepTrak%20100-Global_Report,%20Reputation%20Institute.pdf.

6. See Apple Supplier Responsibility: 2011 Progress Report. Available at http://images.apple.com/supplierresponsibility/pdf/Apple_SR_2011_Progress_Report.pdf.

7. 2011—The Global RepTrak™ 100: The World's Most Reputable Companies. Reputation Institute, p. 6. 2012—The Global RepTrak™ 100: The World's Most Reputable Companies. Reputation Institute, p. 9. At http://www.corporatereputations.ie/news_item.php?id=231; http://www.rankingthebrands.com/PDF/2012%20RepTrak%20100-Global_Report,%20Reputation%20Institute.pdf.

8. See The CSR Africa Daily, at http://csrdaily.csrafrica.net/health/3239-hp-and-ping-collaborate-on-mobile-health-monitoring-to-reduce-malaria.html.

9. See the HP document at http://www.stratos-sts.com/documents/SI_Report_Case_Study_HP.pdf.

10. See the EDF+ Business Web site: http://business.edf.org/projects/innovation-exchange/hunting-energy-treasure-ges-ecomagination-treasure-hunts.

11. See the World Health Organization's International Agency for Research on Cancer, http://www.iarc.fr/en/media-centre/pr/2011/pdfs/pr208_E.pdf.

12. See this press release from the International Telecommunication Union: http://www.itu.int/newsroom/press_releases/2010/06.html.

13. From CTIA, see http://www.ctia.org/media/press/body.cfm/prid/2083.

14. See, from CTIA, http://cellphonehealthfacts.com/faq.html.

15. See World Economic and Social Survey 2011, The Great Green Technological Transformation. United Nations, Department of Economic and Social Affairs (2011) at xxii, 173, 174 (Table v1.3).

16.See list of Current e-Stewards Enterprises at http://e-stewards.org/recycle-responsibly/enterprises/current-e-stewards-enterprises/.

17. See http://www.theclimategroup.org/.

18. http://gesi.org/. 18. http://www.apc.org/en/resources/gesi.

19. See http://www.vestas.com/en/about-vestas/history/1898-1969.aspx, http://www.vestas.com/en/about-vestas/strategy.aspx.

20. See http://gesi.org/portfolio/project/12.

21. Visit the ENLAR Web site at http://iso14000expert.com/fivesteps.html.

22. https://www.globalreporting.org/resourcelibrary/G3-Sustainability-Reporting-Guidelines.pdf.

23. List can be found at http://www.aboutads.info/participating.

REFERENCES

Akpan, Usenobong F., and Godwin E. Akpan. 2012. The contribution of energy consumption to climate change: A feasible policy direction. *International Journal of Energy Economics and Policy* 2 (1): 21–33.

American Association of Advertising Agencies. 2010. Major marketing/media groups launch program to give consumers enhanced control over collection and use of Web viewing data for online behavioral advertising. Press Release (October 4). Available at http://www.aaaa.org/news/press/Pages/100410_aboutads.aspx.

Angwin, Julia. 2010. The Web's new gold mine: Your secrets. *Wall Street Journal*, July 30. Available at http://online.wsj.com/article/SB10001424052748703940904575395073512989404.html.

Baker, Mallen. 2006. So what's the business case for corporate social responsibility? *Business Respect* (101) (August 13). Available at http://www.mallenbaker.net/csr/page.php?Story_ID=1724.

Baker, William E., and James M. Sinkula. 2005. Environmental marketing strategy and firm performance: Effects on new product performance and market share. *Journal of the Academy of Marketing Science* 33: 461–475.

Baucus, Melissa S., and Philip L. Cochran. 2009. An overview of empirical research on ethics in entrepreneurial firms within the United States. *African Journal of Business Ethics*. Available at http://www.ajobe.org/text.asp?2009/4/2/56/63202.

Bhattacharya, C.B., Sankar Sen, and Daniel Korschun. 2008. Using corporate social responsibility to win the war for talent. *Sloan Management Review* 49 (2): 37–44.

Blodgett, John, and Larry Parker. 2008. Global climate change: Three policy perspectives, Congressional Research Service (November 26). Available at http://infousa.state.gov/economy/technology/docs/B2FD2246d01.pdf.

Bloomberg Businessweek. 2004. On the firing line at embattled Merck. December 13. Available at http://www.businessweek.com/magazine/content/04_50/b3912106.htm.

Brown, Darrell, Jesse Dillard, and R. Scott Marshall. 2006. Triple bottom line: A business metaphor for a social construct. Working Paper 200602. Department of Business Economics, Universitat Autònoma de Barcelona.

Campbell, Richard J. 2010. China and the United States—A comparison of green energy programs and policies. Congressional Research Service (June 14).

The Climate Group. 2008. SMART 2020: Enabling the low carbon economy in the information age. Global eSustainability Initiative. Available at http://www.theclimategroup.org/assets/resources/publications/Smart2020Report.pdf.

Coase, Ronald. 1937. The nature of the firm. *Economica* (November).

Cooksey, Kevin, and Delores Kuchina-Musina. 2010. Does size matter? Ethics applied to small businesses compared to large businesses. *Small Business Institute National Conference Proceedings* 34 (1) (Winter).

Cronin, J. Joseph Jr., Jeffrey S. Smith, Mark R. Gleim, Edward Ramirez, and Jennifer Dawn Martinez. 2010. Green marketing strategies: An examination of stakeholders and the opportunities they present. *Journal of the Academy of Marketing Science* 39 (1) (February): 158–174.

Ettenson, Richard, and Jonathan Knowles. 2008. Don't confuse reputation with brand. *MITSloan Management Review* 49 (2) (Winter): 19–21.

Federal Trade Commission. 2010. Protecting consumer privacy in an era of rapid change. Preliminary FTC Staff Report. December. Available at http://ftc.gov/os/2012/03/120326privacyreport.pdf.

Finegold, David, and Allison Moser. 2006. Ethical decision-making in bioscience firms. *Nature biotechnology* 24: 285–290. Available at http://www.nature.com/nbt/journal/v24/n3/full/nbt0306-285.html.

Friedman, Milton. 1999. The social responsibility of business is to increase its profits. In *Ethical issues in business: A philosophic approach*, ed. Thomas Donaldson and Patricia H. Werhane. Upper Saddle River, NJ: Prentice Hall.

Google Public Policy Blog. 2011. Keep your opt-outs. January 24. Available at http://googlepublicpolicy.blogspot.com/2011/01/keep-your-opt-outs.html.

Gross, Grant. 2008. E-waste recycling faces challenges, critics say. *ComputerWorld* (April 30).

Hewlett Packard Global Citizenship. 2010. Custom Report at http://www.unglobalcompact.org/system/attachments/10134/original/HP2010GCR.pdf?1304444102.

Intergovernmental Panel on Climate Change. 2007. Climate change: The physical basis. Working Group I. Cambridge, UK: Cambridge University Press. Available at http://www.ipcc.ch/publications_and_data/publications_ipcc_fourth_assessment_report_wg1_report_the_physical_science_basis.htm. See also http://www.ipcc.ch/publications_and_data/ar4/wg1/en/spmsspm-understanding-and.html.

Kanter, Rosabeth Moss. 2009. *Supercorp: How vanguard companies create innovation, profits, growth and social good.* London: Profile Books.

Ladha, Zahra. 2007. Marketing strategy: Are consumers really influenced by brands when purchasing pharmaceutical products? *Journal of Medical Marketing* 7 (2): 146–151.

Leggett, Jane A. 2009. Climate change: Current issues and policy tools. Congressional Research Service (March 6). Available at http://lugar.senate.gov/services/pdf_crs/energy/Climate_Change_Current_Issues_and_Policy_Tools.pdf.

Meier, Barry. 2011. Sales tactics on implants raise doubts. *New York Times* (May 31). Available at http://www.unmultimedia.org/tv/unifeed/d/17780.html.

Menon, Ajay, and Anil Menon. 1997. Enviropreneurial marketing strategy: The emergence of corporate environmentalism as market strategy. *Journal of Marketing* 61 (January): 51–67.

Minor, Dylan, and John Morgan. 2011. CSR as reputation insurance: Primum non nocere. *California Management Review* 53 (3): 40–59.

O'Marah, Kevin, and Eric Karofsky. 2006. The value of green—A case study of Hewlett-Packard's social and environmental responsibility strategies in the supply chain. AMR Research Report (June).

Parker-Pope, Tara. 2011. Piercing the fog around cellphones and cancer. *New York Times*, June 7.

Phills Jr., James A., Kriss Deiglmeier, and Dale T. Miller. 2008. Rediscovering social innovation. *Stanford Social Innovation Review* (Fall): 34–43.

Piercy, Nigel F., and Nikala Lane. 2009. Corporate social responsibility: Impacts on strategic marketing and customer value. *Marketing Review* 9 (4): 335–360.

Porter, Michael E., and Mark R. Kramer. 2002. The competitive advantage of corporate philanthropy. *Harvard Business Review* (December).

———. 2011. The big idea: Creating shared value. *Harvard Business Review* (January).

Puckett, Jim. 2006. High-tech's dirty little secret: The economics and ethics of the electronic waste trade. In *Challenging the chip: Labor rights and environmental justice in the global electronic industry*, ed. Ted Smith, David A. Sonnenfeld, and David Naguib Pellow. Philadelphia: Temple University Press. Available at http://www.iswa.org/uploads/tx_iswaknowledgebase/2a_-_1100_-_P_-_Puckett.pdf.

Ramstack, Tom. 2011. Environmentalists seek to keep e-waste in North America. All Headline News (February 17). Available at http://ban.org/ban_news/2011/110217_seek_to_keep_ewaste.html.

Reddall, Braden. 2010. Vestas will not chase market share at any price. Reuters, September 1. At http://www.reuters.com/article/2010/09/01/us-vestas-idUSTRE6807MA20100901.

Ruane, Kathleen Ann. 2009. Privacy law and online advertising. *Congressional Research Service* (January 16).

Schramm, Carl J. 2002. The high price of low ethics: How corruption imperils American entrepreneurship and democracy. Speech delivered at the University of Rochester (October 12). Available at http://sites.kauffman.org/pdf/schramm_ethics_10_12_02.pdf.

Sherlock, Molly F., and Donald J. Marples. 2010. Energy tax policy: Issues in the 111th Congress. Congressional Research Service (September 20).

United Nations. 1987. *Report of the World Commission on Environment and Development: Our common future.* New York: United Nations World Commission on Environment and Development. Available at http://www.un-documents.net/wced-ocf.htm.

United Nations Department of Economic and Social Affairs. 2011. *World Economic and Social Survey 2011: The great green technological transformation.* New York: United Nations. Available at http://www.un.org/en/development/desa/policy/wess/wess_current/2011wess.pdf.

United States General Accounting Office. 2008. Electronic waste: EPA needs to better control harmful U.S. exports through stronger enforcement and more comprehensive regulation (August). Washington, DC: U.S. GAO.

Varadarajan, P. Rajan. 1992. Marketing's contribution to the strategy dialogue: The view from a different looking glass. *Journal of the Academy of Marketing Science* 20 (Fall): 335–344.

Wagner, T., R.J. Lutz, and B.A. Weitz. 2009. Corporate hypocrisy: Overcoming the threat of inconsistent corporate social responsibility perceptions. *Journal of Marketing* 73: 77–91.

Werther, William, and David Chandler. 2005. Strategic corporate social responsibility as global brand insurance. *Business Horizons* 48 (4): 317–324.

Wiser, Ryan, and Steven Pickle. 1997. Green marketing, renewables, and free riders: Increasing customer demand for a public good. Report funded by Assistant Secretary of Energy Efficiency and Renewable Energy, U.S. Department of Energy, September. Available at http://eande.lbl.gov/ea/emp/reports/40632.pdf.

8 Global Dimensions of High-Tech Firm Strategies

The Case of the Pharmaceutical Industry

Chapter Overview

This chapter focuses on the rise of the emerging-country pharmaceutical industry in the context of the 1994 Agreement on Trade-Related Aspects of Intellectual Property Protection (TRIPs), the changing structure of pharmaceutical innovation, and concern over the high cost of prescription drugs. It addresses two related issues rather broadly: (1) the strategic response of emerging-country pharmaceutical companies to the new patent regime that recognizes and enforces product patents, with a focus on India; and (2) its implications for the developed world's multinational enterprises' (MNEs) strategies—with respect to collaborations, offshore outsourcing (i.e.,sourcing of goods and services from third parties abroad) and pricing, for example—vis-à-vis the developing-country markets and the ultimate threat to their own home markets. Thus the chapter presents an integrated view of marketing, economic, and legal issues concerning intellectual property rights (IPRs) in the context of the global pharmaceutical industry, which is also referred to as the biopharmaceutical industry. These issues are considered against the backdrop of continuing public policy debates concerning health.

Introduction

The global pharmaceutical industry stands at the center of the *health of nations*—rich and poor alike. Drugs are often substitutes for other more costly means of treatment such as surgery. The innovation of new drugs and their rapid diffusion at affordable prices has been a major source of the phenomenal increase in longevity of the human lifespan over the last hundred years. However, even in the wealthy market economies like the United States, there has always been a tension between the need to preserve incentives for innovation by granting temporary monopoly power to the innovation of new drugs and the need to keep the price of such drugs affordable. The rapid rise of the pharmaceutical industry, specializing in the production and marketing of generic drugs worldwide and especially in the emerging economies such as India, Brazil, and China, is, in part, driven by this tension.

Although a handful of developed countries remain the source of innovation in the global pharmaceutical industry, local firms specializing in the production and marketing of generic drugs in emerging countries have come to first dominate their domestic markets and then challenge the developed world's multinational enterprises (MNEs) in their home markets. Unless otherwise specified, in this chapter, the term MNE refers to the *developed* world's multinational enterprises.

This chapter examines the institutional structure of the global pharmaceutical industry, key players in the industry, R&D funding, TRIPs from the perspective of the developed-country MNEs versus the developing countries, and strategic options available to each, as well as public policy options that could help mitigate the difficult tradeoffs between incentives for innovation and affordable drug prices.

INSTITUTIONAL STRUCTURE OF THE GLOBAL PHARMACEUTICAL INDUSTRY

Figure 8.1 identifies key institutions and players in the global pharmaceutical industry. The shaded circles in the top portion of the figure represent key institutions/players who fund and/or perform much of the world's R&D. The bulk of basic research and much of applied research is funded by governments and performed by universities and their own laboratories. Nonprofit organizations like the Bill & Melinda Gates Foundation provide additional funding of research to universities aimed at developing vaccines for diseases like malaria that are common in the developing world. Integrated MNEs and specialized biotech firms—the latter receiving significant financial support from venture capitalists—fund and perform much of the world's *development* of new drugs. They are the source of much of the biopharma innovation in the world. The firms that produce and market generic drugs in developed and developing countries are significant players in the global biopharma market. Historically, they did little or no innovation, entered the market after the expiration of innovator firms' patents in the developed world, and competed on the basis of cut-throat price competition.

The industry is highly regulated the world over. Regulation involves, for the most part, an elaborate process of approving new drugs, postmarket surveillance, advertising and other forms of promotion, and prices in most countries. Customers of the pharmaceutical industry, including physicians, patients, health maintenance organizations (HMOs), pharmacies, hospitals, governments, and businesses, affect and are affected by the regulatory system in each country. The industry's institutional structure also includes a global network of alliances and a supplier network for R&D, production, and marketing. Protection of intellectual property rights during the patent life of drugs is central to innovation as well the success of innovator firms. The TRIPs, discussed

Figure 8.1 **Key Institutions and Players in the Pharmaceutical Industry**

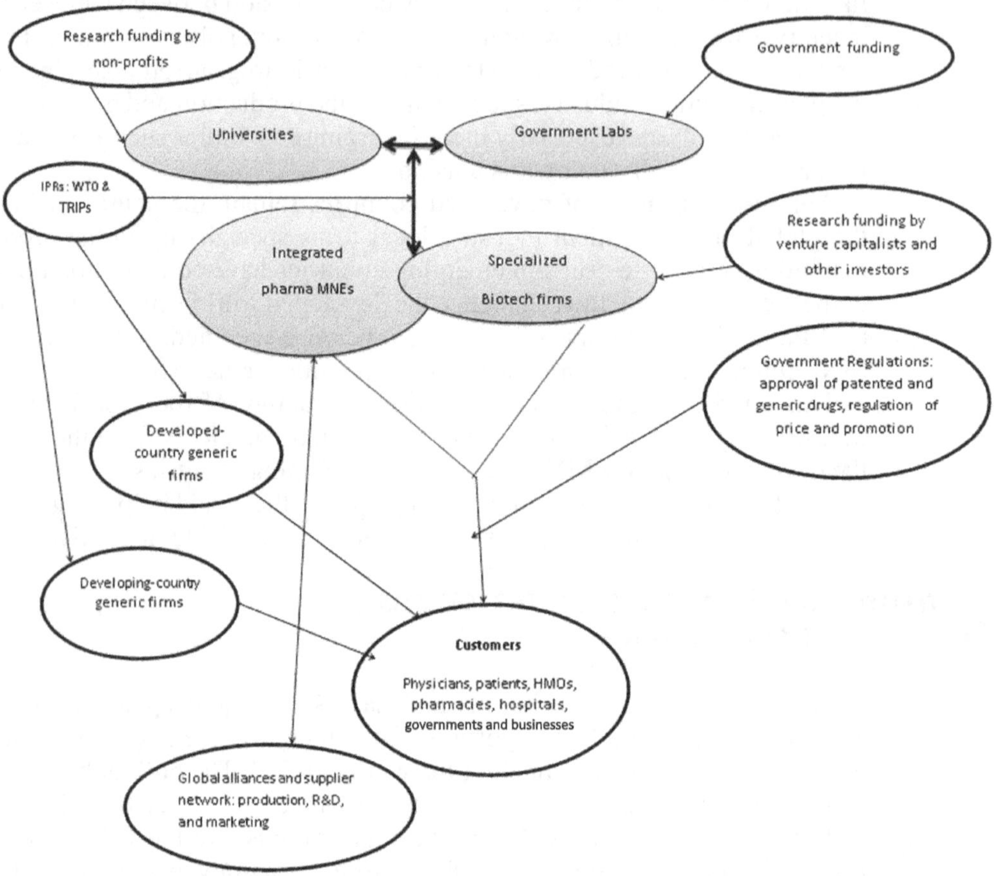

Sources: Adapted from Pisano (2006, p. 81, Fig. 5-1); and Ghauri and Rao (2009, p. 8, Fig. 2).

below, represent an attempt to extend the developed world's IPRs to the developing countries as part of the global trading system.

THE TRIPS AGREEMENT

Few issues in recent years have generated and continue to generate more North-South economic debate and controversy than the 1994 Agreement on Trade-Related Aspects of Intellectual Property Rights. TRIPs is now a key part of the trading system under the World Trade Organization (WTO) framework.

Perhaps the most comprehensive international agreement on intellectual property rights ever negotiated, the TRIPs agreement involves more than the narrow issue of trade flows between nations. Indeed, the term "trade-related" is something of a misnomer in the sense that the agreement requires sweeping changes in the IPR regimes of many developing countries with important implications for *economic development* that go beyond the narrow trade issues. The agreement has implications for the benefits of MNE innovation, funding and location of R&D, MNE linkages with the public sector, academic institutions and foundations, technology transfer, and spillovers.

Harmonization of legal instruments governing intellectual property protection is viewed as a response to the appropriability problem in the TRIPs agreement. Indeed, Leebron (1996), who provided an extensive analysis and critique of harmonization claims, suggests that for the first time, the concept is firmly incorporated in the WTO framework for IPRs as the norm for international economic relations. It should be noted that the concept of harmonization adopted in the TRIPs agreement insists that the law and enforcement mechanisms governing IPR protection in the developing countries conform to a higher standard, namely, that of the developed world. Note that the TRIPs agreement sought not merely the "sameness" of laws across borders. Rather, TRIPs sought "sameness" of laws in accordance with the developed world's standards. The underlying assumption is that if the developing world adopts stronger IPR regimes, appropriability of the fruits of the developed world's innovative effort would be enhanced.

In the context of pharmaceutical MNEs, the implications of the TRIPs agreement go to the heart of the health of nations via its potential impact on innovation and the price of patented drugs. This is because IPRs in the form of patents provide a major incentive for innovation in the pharmaceutical industry and are an important determinant in the pricing of medicines. Medicines with patents, but without a therapeutic equivalent, enjoy market exclusivity—twenty years—within the jurisdiction of the patent and they are priced according to what the market bears. In the developed world, legal competition during the life of a drug's patent is limited to a therapeutic equivalent in a different brand. Prior to TRIPs, legal competition from generics before patent expiration existed in the developing countries that did not recognize patents for pharmaceutical products, like India.

The debate between the developed world and the developing world concerning the application of TRIPs to the developing world's pharmaceutical industry, which thrived on rapid and low-cost imitations prior to patent expiration, was particularly acrimonious, running the gamut of broad issues concerning strategy, structure, and performance of the developed world's MNEs. The larger issue for the developing world is the potential negative impact of TRIP argument on drug prices and the health and welfare of its population. This is despite the TRIP agreements, which allows the least-developed countries with no domestic production to import cheap drugs subject to certain conditions aimed at preventing abuse. The issue for the developed world, which is the primary source of pharmaceutical innovation, is to protect its firms from foreign imitations and minimize the appropriability problem. Note that, according to estimates by Lichtenberg (2003) and Scherer (2004), private rate of return on pharmaceutical innovation is only one-fifth of the return to society, which cannot be captured by the innovating firms.

TRIPs Provisions: The Developed World's MNEs Versus the Developing World

Prior to TRIPs, pharmaceutical products had no patent protection in many developing countries. They are now patentable, at standards prevailing in the developed countries prior to the TRIPs, and applied uniformly, albeit with a delay in case of the poorest countries. The delay in application of the TRIPs provisions to the poorest countries—due to a provision for market exclusivity during the transitional period—has been extended to 2016 (Braga 1995; International Federation of Pharmaceutical Manufacturers Association 1995; WTO 2001). Overall, TRIPs provisions and standards for patents have succeeded, at least on paper. They were further reaffirmed in the 2001 Doha Declaration (WTO 2001). Still, the issue of whether the current patent system serves as a guide to effective public policy is by no means settled. Gallini (2002) reviewed the vast literature on this subject and concluded that patents may be too blunt an instrument to apply to all technologies in all environments. More recently, Lanjouw (2005), utilizing 68-country 20-year drug launch data, concluded that for low- and middle-income countries the evidence that stronger patent protection leads to greater access to new drugs is mixed.

ISSUES FOR THE DEVELOPED WORLD'S MNES

MNEs invest heavily in R&D and it is risky. During the decade preceding 2010, R&D intensity (i.e., the R&D to sales ratio) of U.S. MNEs averaged 16.5 percent (PhRMA 2010). Indeed, judging from the U.S. patent data reported by Lanjouw (2003), it is clear that pharmaceutical innovation is the stuff of the developed world's MNEs. Just eight developed countries—the United States, Japan, Germany, United Kingdom, France, Switzerland, Italy, and Canada—accounted for 91 percent of U.S. pharmaceutical patents between 1985 and 1995. Confirming Lanjouw's finding, the United States alone accounted for half of the total in more recent data reported by Friedman (2010), whose methodology for assigning patents to country or region of invention considers multiple geographically dispersed inventors. Of the 1,400 first-year patents granted to drugs approved between 2001 and 2009, 60 percent of inventors were from the United States and 31.5 percent of inventors were from just seven other developed countries (United Kingdom, Japan, Germany, Sweden, France, Switzerland, and Belgium; see Table 8.1).

The drug discovery process has become more complex, and it takes 10 to 15 years to develop a new drug. The average cost of researching, developing, and introducing a new drug in the United States in 2005 was about $1.3 billion, almost ten times the cost in the mid-1970s (DiMasi, Hansen, and Grabowski 2003; DiMasi and Grabowski 2007).

Only three of ten marketed drugs produce enough revenues to cover or exceed average R&D costs (Grabowski, Vernon, and DiMasi 2002). The effective patent life in the United States averages about ten years. As shown in Figure 8.2, generic competition in the domestic market has increased significantly as has inter-brand competition. Between 1990 and 2010, the generics' share of U.S. prescription drug market has more than doubled, from 33 percent to 78 percent (PhRMA 2010).

In the absence of patent protection for pharmaceutical patents, developing-country generics, unlike the U.S. generics, did not have to wait for patents of the MNEs to expire before imitation began. Countries such as Argentina, Brazil, India, and Mexico developed a thriving generic drug industry. The drugs were marketed at home as well

Table 8.1

Percentage of U.S. Pharmaceutical Patents, by Nationality
(average 2001–2009 annual data)

Nationality	Patents (%) National	Patents (%) Cumulative
U.S.	64.26	64.26
UK	8.65	73.91
Japan	7.78	81.69
Germany	4.69	86.38
Sweden	4.77	91.15
France	3.10	94.25

Source: Authors' calculation based on data from Friedman (2010) for 1,400 first-year U.S patents granted to drugs approved between 2001 and 2009.

Figure 8.2 **Generics' Share of U.S. Prescription Drug Market, Selected Years**

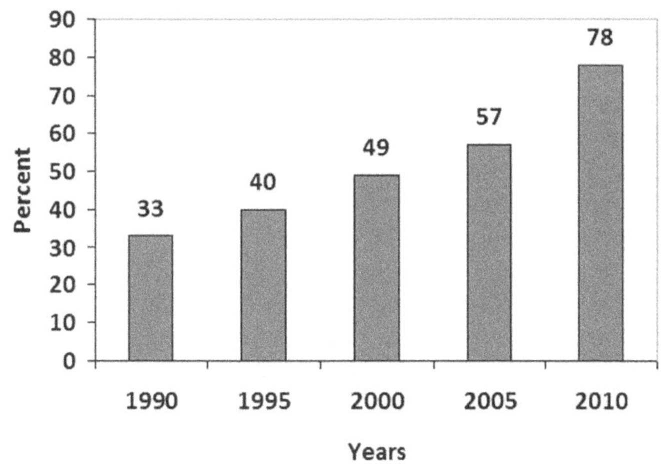

Source: PhRMA (2011).

as abroad, mostly to other developing countries, but increasingly to developed countries that include Switzerland, Germany, Canada, and the United States (exports to the United States occur via Canada) at a fraction of the developed world's prices.

ISSUES CONCERNING THE DEVELOPING WORLD

Despite the Doha Declaration, which is viewed as a victory for the developed world's pharmaceutical MNEs, widespread criticism of patent rights continues not only in the developing world but also among many scholars in the developed world (see Stiglitz 2003, for example). The high incidence of HIV/AIDS in the developing world, combined with unaffordable prices of drugs to treat it, have only accentuated the North-South debate over TRIPs.

The first and foremost issue for countries in the developing world is the impact of TRIPs on drug prices. As Lanjouw (2003) suggests, because health is such a basic need, higher prices due to patents impose greater hardship on developing countries than on developed countries. Under TRIPs countries like China and India, which were in the $400 to $500 range of GDP per capita (in 1995 U.S. dollars) had to adopt pharmaceutical product patents, whereas Norway with $30,389 GDP per capita had introduced patent protection as recently as 1992 (Table 8.2). Moreover, data analyzed by Lichtenberg (2010) on 6,500 drugs sold in 38 countries during first-quarter 1999 and third-quarter 2008 show that drug prices in developing countries like India, China, Egypt, Indonesia, and Malaysia are in the range of 20 to 40 percent of U.S. prices.

Kremer (2002) suggests that the characteristics of the pharmaceutical market in the developing world differ in many ways from those in the developed world. For one, their markets are small. In 2009, per capita public and private health expenditures in the United States were $7,960 compared with $124 on the basis of purchasing power parity in India, $347 in China, and $921 in Brazil. In many poor countries, pharmaceutical firms do not even take out patents, as they face a different disease environment. Low-income countries account for 36 percent of deaths from infectious and parasitic diseases; lower-middle-income countries 54 percent; middle-income countries 8 percent; and high-income countries 2 percent (World Health Organization [hereafer WHO] 2008). Still, many diseases like cancer and heart disease affect both the developed and the developing world. Moreover, the developing countries, in general, have weak health-care systems due to the shortage of trained physicians, inefficient procurement and distribution, self-prescription, and other flaws.

Indeed, the ongoing bitter debate over the price of antiretroviral drugs to treat HIV/AIDS has as much to do with the price of such drugs as with weak health-care systems. The issue for poorer countries is that a vast majority of their populations cannot afford any drugs, much less antiretrovirals. As a result, the coverage of antiretroviral therapy in Africa, where the estimated need is greatest, was only 50 percent in 2010 (WHO 2011, 2012).

Another issue has to do with the location of pharmaceutical MNE R&D in the developing countries. Judging from the U.S. data (PhRMA 2010), very little, if any, of the pharmaceutical MNE R&D is located in the developing world. As shown in Table 8.3, in 2010, India, China, Brazil, and Mexico combined accounted for 0.7 percent of the MNE worldwide R&D, or less than one-fifth of their share in sales. An extraordinarily high proportion of the MNE R&D abroad (96.7 percent) was located in just four countries/regions: United States (80.2 percent), Western Europe (14.1 percent), Japan (1.4 percent), and Canada (1.0 percent).

International business scholars stress the quality of local science and large markets for drugs as two important factors influencing the choice of a country for pharmaceutical R&D (Burstall and Dunning 1985). This view appears to hold for some developed countries, such as the United Kingdom, Germany, and Japan. However, much of the developing world has neither the quality of science nor the large market. Very little of R&D on drugs for diseases common to the developing world is done by the MNEs in either the developed world or the developing world. Lanjouw (1997) notes that almost all R&D for such diseases comes from either internationally funded

Table 8.2

Development Level and Adoption of Pharmaceutical Product Patents, Selected Countries

Country	Year of Adoption	GDP per capita (in 1995 USD)
Developed Countries		
Japan	1976	$24,043
Switzerland	1977	36,965
Italy	1978	13,465
Sweden	1978	21,896
Canada	1983	16,296
Denmark	1983	28,010
Spain	1992	14,430
Norway	1992	30,389
Developing Countries		
China	1992/93	424*
Brazil	1996	4,482
Argentina	2000	8,100
Uruguay	2001	6,208
Egypt	2002	1,191
India	2005	450

Source: Adapted from Lanjouw (2003, p. 96).
*GDP is for 1992.

Table 8.3

Share of U.S. Pharmaceutical MNE R&D and Sales by Geographical Area, 2010

Country/Area	R&D (%)	Sales (%)
United States	80.2	63.4
Western Europe	14.1	16.2
Japan	1.4	4.6
Canada	1.0	2.3
Central and Eastern Europe	0.7	3.7
Australia	0.4	1.4
China	0.3	1.1
Brazil	0.2	1.4
Mexico	0.1	0.9
India	0.1	0.4
Other	1.5	4.6
Total	100.0	100.0

Source: PhRMA (2011).

organizations or the military in the developed countries. She reports data on twenty such diseases (e.g., malaria and polio) for which virtually 100 percent of the burden falls on low- and middle-income countries.

Moreover, there is a growing and strong body of literature on health and development that treats health as a determinant of economic growth and welfare in a number of rather obvious but important ways (see Bloom, Canning, and Jamison 2004, for example). Workers who are healthy are more productive than workers who are not; poor health reduces gross domestic product (GDP) per capita by reducing both labor productivity and the relative size of the labor force; rising longevity leads to planning for retirement and incentive for the current generation to save; healthier children have higher rates of school attendance; and longer life span makes investment in education more attractive. Bloom, Canning, and Malaney (2000) provide compelling evidence along these lines on the leading role played by health improvements in the East Asian economic miracle. Cutler, Deaton, and Lleras-Muney (2006, p. 115), in a study of the determinants of mortality within countries, corroborated these findings. The authors conclude that "much of the link between income and health is a result of the latter causing the former, rather than the reverse."

Still another issue for the developing world, especially for countries like India, had to do with the adverse effect of TRIPs on the thriving generic drug industry, which was built almost entirely on early and cheap imitation of new and patented drugs produced in the developed world. Recall that most developing countries did not recognize pharmaceutical product patents prior to TRIPs.

STRATEGIC RESPONSE OF THE EMERGING-COUNTRY FIRMS

COMPETE AND COLLABORATE

Sampath (2006a) examined this issue at length with the aid of data derived from a survey of 103 Indian pharmaceutical firms categorized into three groups: group 1 consists of 31 large-scale firms, both Indian and MNE-held; group 2 is 27 medium-sized operators who are either generic producers or specialists in niche areas of contract research; and group 3 is 45 small-scale units that manufacture drugs for the bigger firms within India. Not surprisingly, Sampath's analysis of the survey data suggests that the response of Indian firms to the TRIPs agreement varies with the group. A summary of emerging strategies is presented in Table 8.4.

All these groups are driven by the need to close gaps, post-2005, in their respective competitive positions as well as leverage their strengths. Note that firms in all three groups are global players, albeit to varying degrees. The group 1 companies are moving or have moved toward the traditional pharma MNE model with a high degree of backward integration into R&D as well as forward integration into marketing and distribution. Potentially, it is a high-risk and high-return strategy. The firms in group 2 will continue to leverage their skills in reverse engineering focused now on off-patent drugs without violating the TRIPs agreement. The risk-return profile of these companies tends to be in the medium range. By contrast, group 3 companies, which specialize in manufacturing for group 1 and group 2 companies, are in the low-risk and low-return range. However, it is important to note that companies in all three groups have become or could become global players. This view is supported by the group-level evidence on local and foreign collaborative links presented by Sampath (2006b). While foreign collaborative links are greater for group 1 firms, group 2 runs a close second to group 1 and group 3 to group 2. Moreover, as shown in Table 8.5, the specific competitive and cooperative strategies available to Indian firms—and increasingly adopted—run the

Table 8.4

Emerging Strategic Focus of Indian Firms: A Categorization

Firm group	Drivers	Strategic focus	Relative risk and return
Group 1—large size (e.g., Ranbaxy, Dr. Reddy's Labs)	Need to strengthen product portfolios to compete in global market	Greater investment into R&D, higher innovation in generics, new products and processes, and bulk drugs	High
Group 2—medium size (e.g., Matrix Labs, Ajantha Pharma)	Need to strengthen competitive advantages to make use of contract research and manufacturing services contract research and manufacturing services (CRAM) opportunities	Active supply of off-patent generics to the semi-regulated and unregulated markets; establish themselves as niche players for contract research	Medium
Group 3—small size (e.g., Radico Remedies)	Survival in light of India's full-fledged TRIPs compliance	Upgrading facilities to continue manufacturing for group 1 and group 2 companies	Low

Source: Adapted from Sampath (2006a).
Note: Last column of the table is authors' addition.

Table 8.5

Strategic Options for Indian Companies

Option to compete	Option to cooperate
• Market and sell plain vanilla and specialty generics	• Provide contract manufacturing for MNCs
• Develop low-risk NDAs	• Supply APIs to MNCs
• Develop follow-on biologics	• Partner with MNCs for their sales channels
• Challenge IPRs on regulated markets	• Provide clinical outsourcing for MNCs
• Invest in R&D for proprietary NCEs	• R&D collaboration

Source: Grace (2004).

entire gamut from new chemical entities to novel drug-delivery systems to in-licensing arrangements to collaborative R&D to contract research.

Innovation

The top-tier pharmaceutical firms in the emerging economies are investing in innovation with the aim of becoming global players in the industry. Between 2004 and 2009, R&D expenditures of seven firms ranked among the top ten Indian firms by sales in 2009 doubled to $371 million, and R&D intensity (the ratio of R&D to sales) of these firms increased by about 80 percent to 8.9 percent during the same period compared to the industry average of about 5.0 percent (Mitra 2006; Bruche 2011; Joseph 2011). According to the WHO-INTECH survey results presented by Sampath (2006b), the emerging R&D strategies by the Indian firms appear to focus on collaborative research, clinical trails, and generics. This is true of all three groups, albeit to varying degrees.

As might be expected, patenting is a relatively new phenomenon in the Indian pharmaceutical sector. Cipla, Ranbaxy, Dr. Reddy's Labs, and Matrix Labs are among the leading patenters. These companies engage in positive patenting (i.e., to secure worldwide intellectual property rights on their innovations) as well as defensive patenting (i.e., to prevent others from obstructing their R&D activities). Morel et al. (2005) provide an interesting statistic concerning India's rank in patenting. When the top 25 countries are ranked in order and analyzed for all U.S. patents issued where at least one inventor is from a given subject country, India ranked third highest after the United States and Japan. They further find that the number of U.S. patents per GDP per capita held by India is 0.912, second only to the United States and Japan. Thus, it appears that TRIPs compliance could, in the longer term, have the beneficial effect of shifting the Indian pharmaceutical industry from mere producers of generics drugs to innovators of new drugs.

OFFSHORE OUTSOURCING PROJECTS FOR MNES

Indian pharmaceutical companies already have significant presence in clinical development—which accounts for about 40 percent of the MNE R&D budget—and, as noted earlier, are increasingly entering into drug discovery, which accounts for about 30 percent of the budget. Overall, Bangalore, Mumbai, and Hyderabad have become the hubs for clinical development, especially clinical data management. It is expected that pharmaceutical offshore outsourcing to India will grow to around $7 billion by 2013 as global firms seek to take advantage of cost and quality offered by Indian firms (Unnikrishnan 2007). Indian companies are increasingly focusing on contract research and manufacturing services (CRAMs). Contract research includes:

- Drug discovery and development services
- Custom syntheses
- Process development
- Preclinical trials
- Clinical research and clinical trials

India becomes all the more attractive for outsourcing because all phases of the pharma business—discovery, development, manufacturing, and marketing—are increasingly linked with information technology (IT) services. For example, GlaxoSmithKline is seeking not only a full range of services related to discovery, development, and contract manufacturing but also IT application development and infrastructure.

STRATEGY OPTIONS FOR THE DEVELOPED-COUNTRY MNES

PRICING STRATEGIES

Differential Pricing

A number of scholars have advocated *differential pricing* or price discrimination as a way around the tradeoff between incentives for innovation and affordable drug prices for the developing countries (Kremer 2002; Lanjouw 2003; Hellerstein 2003; Barton 2004; Vachani and Smith 2004; Danzon 1997, 2000; Rao 1998, for example). The basic concept of price discrimination and the way marketers implement it was fully discussed in chapter 3. Price discrimination—in the context of international marketing of pharmaceuticals—includes selling the same patented drug at different prices in different countries depending on what the market bears—lower prices in the developing countries where demand is relatively price-elastic, and higher prices in the developed countries where demand is relatively price-inelastic.

While a differential pricing proposal could become part of the solution, it is not without significant problems. One such problem is the potential for the emergence of gray markets via re-exports from the developing countries to the developed countries, undermining the ability of pharmaceutical firms to price-discriminate. Another problem is political. Even if the gray market can be minimized with tight border controls, an MNE selling the same patented prescription drug at the monopoly price in its domestic market and, say, at or near the marginal cost in the developing countries is likely to be politically unpopular at home. This is despite the fact that the practice of price discrimination in the domestic markets by the MNEs is pervasive. For example, the U.S. government and hospital and health maintenance organizations routinely negotiate lower prices for prescription drugs (Berndt 2002).

Disguised Price Discrimination

More important, MNEs often practice disguised price discrimination (akin to third degree price discrimination) across countries. Disguised price discrimination involves marketing the same product at different prices, across countries, but under different names, packaging, and the like. For example, at the time of patent expiration, Glaxo sold its antiulcer drug Zantac at about $2.80 for 300 mg in the United States and at the same time sold the same drug at less than 6 cents under the name of Zintec in some countries (Lanjouw 2003).

Another case in point is the recent introduction of an antimalarial pill by Sanofi-Aventi called ASAQ that combines artemisinin, invented in China, with amodiaquine, an older drug, at a price of $1 for poor adults and less than 50 cents for poor children in developing countries. Sanofi has decided not to seek patents. It sells at cost to international agencies like the World Health Organization. With creative packaging and monitoring of pharmacies, Sanofi will provide a branded version, coarsucam, targeted at the high-income segment, which will be sold at three or four times the price paid by the poor customers. Packaging of the cheaper version will be different and there will be monitoring of pharmacies to ensure that they are not selling the cheaper version at a higher price (McNeil Jr. 2007).

The more that price discrimination can be disguised in terms of packaging, labeling, convenience, and other factors, the greater the likelihood that it will be successful. Note that multiple presentations in which pharmaceutical products are available—dosage forms (e.g., capsule, tablet, syrup), strength (e.g., 50 milligrams, 100 milligrams), and packet size (e.g., 50 tablet bottle, 100 tablet bottle or customized), collectively referred to as stock-keeping units (SKUs)—combined with greater adaptation of the product itself to the needs of particular countries offer many opportunities for effective price discrimination.

Reference Pricing

Notwithstanding the patent system, most developed as well as developing countries have cost containment/price control measures in one form or another (Organization for Economic Co-operation and Development 2006). One form of cost containment that is popular in some European counties (e.g., Belgium, Norway, and the Netherlands) is external *reference pricing*, whereby a price in one country is capped at the lowest price for the same drug in a set of other countries. External reference pricing, like re-exports, dilutes the ability of MNEs to engage in differential pricing. This means that the inability to price discriminate could result in launch delays or no launch of new drugs in the developing countries (Danzon 2006).

Socially Responsible Pricing

While viewing price discrimination across markets as a beneficial instrument that keeps prices of essential drugs lower than they would be otherwise for the poor countries, Vachani and Smith (2004, p. 118) propose the concept of *socially responsible pricing* (SRP), which is "pricing that attempts to sustain or enhance *social welfare*" (emphasis added). Bull and McEvoy (1999) have argued that the MNE's ethical capability, that is, its ability to respond effectively to ethical issues, is a source of competitive advantage. Given the extensive regulation and public attention focused on the pharmaceutical MNEs, reputation benefits derived from SRP could lead to economic benefits in the form of easier access to the developing world's markets. The general issue of the relationship between a firm's ethical behavior and potential reputational benefits it derives from it was discussed at length in chapter 7. It should be noted that SRP suffers from some of the same problems (e.g., re-exports and political backlash in the home country) as well as straightforward price discrimination across countries.

MARKETING ONE'S OWN AUTHORIZED GENERIC VERSIONS

Another pricing strategy for the developed-country MNEs is to offer their own generic version in the global markets. At least two recent cases illustrate this strategy: the case of Pfizer's top-selling cholesterol drug, Lipitor, and the case of Roche's cancer drugs Herceptin and Mab Thera. After losing its patent on Lipitor—the first drug to exceed \$10 billion annual sales—Pfizer entered into an agreement with Watson Pharmaceuticals to market a generic version authorized and manufactured by Pfizer and negotiated deals with insurers and pharmacy benefit managers to match or beat the generic alternatives (Wilson 2011).

In the case of brand-name cancer drugs, Herceptin and Mab Thera, Roche entered into agreement with an Indian partner, Emcure Pharmaceuticals, to rename and package cheaper versions. This action by Roche came after the Indian government denied Bayer the exclusive rights to produce the kidney cancer drug Nexavar and licensed a local company to make a cheaper generic version on the grounds that many Indians could not otherwise afford the life-saving drug. Note that there is an escape clause in the TRIPs that allows national governments, under certain conditions, to require compulsory licensing in the interest of national health (Kollewe 2012).

LICENSING AND SALE OF TECHNOLOGIES

The essence of a market for technology is what Pisano (2006b) calls *monetization of intellectual property*. "That is, know-how itself has become a separable asset that is traded, valued and appropriated" (p. 9). As discussed in chapter 1, licensing has become the chief means by which technology transactions take place. It has also become one of the rapidly growing means of monetizing intellectual property. Out-licensing of pharmaceutical product patents to the developing-country firms is a strategic option. Outright sale of patents is another means by which firms monetize intellectual property. Note that both licensing and sale options could accentuate the appropriability problem because of spillover of proprietary technology to potential competitors with or without legal protections. For this reason, traditional vertically integrated firms—that is, firms with R&D, production, and marketing under one roof—find it unattractive to out-license or sell, at least until the product reaches late maturity to declining stage in the product life cycle. Indeed, in-licensing, that is, licensing from other MNEs as well as from the emerging-country firms, is more common. However, the U.S. firm Napo Pharmaceuticals' out-licensing deal involving an antidiarrheal molecule with the Indian firm Glenmark could serve as a model for many such deals. Note that IPRs over the composition of matter of the compound are with Napo. Glenmark has been given the license to develop and commercialize the drug in 140 countries including India, but excluding the United States, Europe, China, and Japan. In addition, the deal contained a contract manufacturing provision by which Glenmark would exclusively supply Napo's global requirements for the manufacturing and sale of the drug. Moreover, pure biotech firms whose job it is to produce innovations without production and marketing capabilities may find out-licensing relatively more attractive.

OPEN INNOVATION

As discussed in chapter 1, several of the trends noted—the growing importance of private-public nonprofit linkages coupled with an increasing trend toward alliances and licensing—may be viewed as indicators of the emerging open innovation (OI) paradigm, which has received much attention in recent years (Chesbrough 2003). Merck's Gene Index story (discussed in chapter 1) involving collaboration with Washington University—and making what would otherwise be considered proprietary results public—illustrates the potential benefits and costs of adopting the OI model of innovation. The OI model is designed, among other things, to increase the quantity and quality of the innovation pie, from which the innovating firm and its collaborators derive greater benefits than would be the case otherwise. Thus, large pharmaceutical firms with strong R&D and downstream marketing assets may gain more by adopting the collaborative network-driven OI model even if it meant that a significant amount of proprietary knowledge spills over into the public domain. Therefore, the advocates of the OI model suggest that firms adopting the model may find it is compatible with a high degree of appropriability even as it dilutes their IPRs. Moreover, to the extent that stronger IPRs in the developing world encourage greater technology linkages with the MNEs, the outcome would benefit rich and poor countries alike.

PUBLIC-PRIVATE LINKAGES

The conventional view that pharmaceutical research is a linear process in which the basic research discoveries funded by the public sector are turned into drugs has been questioned by several scholars. Pharmaceutical data on the pattern of co-authorship between private sector firms and "public" institutions (hospitals, universities, and the National Institutes of Health, or NIH, for example), documented by Cockburn and Henderson (1996), show strong linkages between private sector research and "upstream" public sector research. More important, the results support the hypothesis that the linkage with public sector basic research activity is an important determinant of the productivity of downstream private sector research. Yamin and Otto (2004) examined the effect of inter-organizational and intra-organizational knowledge flows on innovative performance of pharmaceutical MNEs and confirmed the importance of public-private linkages, especially the academic–private sector linkages. Their results strongly support the hypothesis that external knowledge flows from academia have a significantly positive influence on innovative performance of the private sector.

Although collaborative research relationships in the United States between university faculty and industry were common as far back as the turn of the century, the 1980 Bayh-Dole Patent and Trademark Amendments Act gave additional incentives for U.S. universities to translate basic research discoveries arising from federally funded projects into commercial products by granting them patents (Mowery et al. 2001). The logic behind the Bayh-Dole Act was that companies needed exclusive patent rights to speed up the development and commercialization of publicly funded university research. In 2009, about 52 percent of U.S. academic R&D expenditures went for medical and biological sciences, a significantly greater proportion compared to other developed countries (National Science Board 2012, Table 5-5).

University patenting and licensing began to grow rapidly after 1980 in the wake of Bayh-Dole Act. Gross university licensing income grew more than ninefold between 1991 and 2010, from about $ 0.2 billion to well over $1.8 billion (Sampat 2006; Association of University of Technology Managers [hereafter AUTM] 2010). However, much of this activity is confined to a few research universities. Just one process patent issued in 1980 on a fundamental technique for the creation of genetically engineering microorganisms produced—over its seventeen-year life—$255 million for Stanford University and the University of California (Kenney and Patton 2005). Moreover, universities are increasingly involved in startup companies, especially in life sciences research, headed by the faculty who also run academic research groups. Over a period of three decades since the passage of the Bayh-Dole Act in 1980, there has been a thirteen-fold increase in the number of patents *issued* to research universities in the United States (Sampat

2006; AUTM 2010). Note that while research at many of these centers is focused on specialized diseases and not undertaken by the private sector, their infrastructure and capabilities are commercial in nature (Pollack 2003; Mehta 2004).

Several questions arise concerning academic–private sector research linkages for developed as well as developing countries. What are the consequences of privatization of pharmaceutical-related academic research results in the developed countries for future innovation in the developing countries? For example, what effect would increased patenting by the universities have on the nature and speed of academic publications? And on access to and cost of even normal scientific knowledge for the developing countries?

THE ROLE OF FOUNDATIONS

Nonprofit foundations fill the gaps in research funding left by the public sector and the for-profit private sector. They often fund research that is scientifically speculative, politically unpopular, and where contribution of foundations to health research is measured less by the dollar amount and more by the areas that otherwise would not get even the seed money. In recent years, a new form of nonprofit foundation/institute funding called *venture philanthropy* has come into being. Venture philanthropy combines the features of not-for-profit entities and for-profit venture capital firms in the sense they take a more active approach by providing management guidance and monitoring the progress against benchmarks (Pisano 2006b). Venture philanthropists typically fund "translational research" involving target identification and validation, early-stage human clinical trials and the like, which are given low priority by the National Institute of Health. The Bill & Melinda Gates Foundation, which funds research on malaria and HIV/AIDS vaccines, is an example of *venture philanthropy*. In 2010, the Gates Foundation paid grants in the amount of about $1.5 billion, mostly for research on diseases affecting developing countries (Bill & Melinda Gates Foundation 2010). During 2007–2008, foundations, charities, and other private sources funded a little over $4 billion, or about 4 percent of total funding, for biomedical research (Dorsey et al. 2010).

PUBLIC POLICY OPTIONS

Given the strong role of government in many aspects of the global pharmaceutical industry—in funding and/or performing basic research, in granting patents, in approving new drugs, in buying drugs for public hospitals, in regulating and/or influencing prices and marketing, and in developing and enforcing the IPRs, for example—public policy is often at the center of the North-South debate over preserving incentives for innovation versus keeping drug prices affordable. We discuss next two public policy options that have been advanced in this context.

DIFFERENTIAL PATENT PROTECTION

Lanjouw (2003) proposes a system of patent protection with the aim of increasing private sector incentives to discover drugs for diseases that predominately affect poor countries. Under this system, MNEs would be discouraged from seeking patents for global diseases (e.g., cancer) in the developing countries, while allowing them to enjoy patent protection (according to the TRIPs) on drugs such as malaria.

There are at least three problems with this proposal. One problem is that a differential patent system leads to differential pricing, whose shortcomings were previously noted. A second problem is that patent protection is greatly undermined when it includes vaccines because of positive externalities associated with vaccine consumption noted earlier. That is, the market for vaccines quickly evaporates with vaccination. Third, even if the MNEs find sufficient incentives to discover drugs for "poor country diseases," poor countries will not benefit unless there is price regulation.

EXPANDED RESEARCH FUNDING: PUSH VERSUS PULL APPROACH

The *push* approach involves paying for research inputs (e.g., NIH funding of academic research and general tax credits to the private sector), whereas the *pull* approach involves paying for research outputs (e.g., grants for development of specific vaccines, say, for AIDS). Each has its advantages and limitations (Kremer 2002; Kremer and Glennerster 2004). The push approach, because it is not aimed at any particular product, is best suited for funding basic research, whereas the pull approach is best suited for development. Publicly funded research as well as research funded by nonprofit and international organizations includes both approaches in the developed countries. However, given the very large social returns to pharmaceutical innovation, the case for expanding the scope and size of funding by public and nonprofit institutions the world over and international organizations is strong.

Arrow (2004, p. 21), writing about the need for new antimalarial drugs and recognizing why it would not pay for pharmaceutical companies to develop the much-needed Artemisia combination therapy, asks: "What incentives can be created to encourage public and private research of these issues?" and answers: "Something beyond the ordinary intellectual property rights seems to be necessary: public sector investment in research and development."

SUMMARY

In this chapter we have presented global dimensions of high-tech firm strategies in the context of the pharmaceutical industry. The chapter examined the institutional structure of the global pharmaceutical industry, key players in the industry, R&D funding, TRIPs from the perspective of developed-country MNEs versus developing countries and strategic options available to each, as well as public policy options that could help mitigate the difficult tradeoffs between incentives for innovation and affordable drug prices.

The institutional structure of the global pharmaceutical industry is complex, with many players beyond the pharmaceutical MNEs. Governments everywhere fund and/or perform most of the basic research and some applied research as well. Governments also regulate the industry in many ways, including the granting of patents, approval of new drugs, controlling and/or influencing prices and promotion. Large, vertically integrated pharmaceutical companies fund and perform much of the developmental work associated with new drugs as well as manufacture and market them. Specialized biotech firms, often populated by star scientists and funded by venture capitalists and other investors, perform research at a more fundamental level of discovery—and through a new route—leading to new drugs.

The developed world's pharmaceutical MNEs invest heavily in R&D—about 17 percent of sales—and only three in ten drugs produce enough revenues to cover or exceed average R&D costs. Effective patent life in the United States averages about twelve years. Innovation in the pharmaceutical industry is highly concentrated in the developed countries. Just eight countries account for 91 percent of innovation in pharmaceuticals. After the patent expires, generic firms enter and engage in cut-throat price competition. Prior to TRIPs, many emerging countries did not recognize pharmaceutical product patents. As a result, emerging-country generics did not have to wait for patents to expire before imitation began. They first controlled the domestic market and eventually began to export to other developing countries as well as the developed-country markets. It was against this background that the developed-country MNEs pushed for TRIPs in order to increase the appropriability of their innovations.

HARMONIZATION OF LEGAL INSTRUMENTS

Legal instruments governing intellectual property protection is viewed as a response to the appropriability problem in the TRIPs agreement, which generated an intense debate between the developed and the developing world. It ran the gamut of broad issues concerning strategy, structure, and performance of the developed world's MNEs. The larger issue for the developing world is the potential negative impact of TRIPs on drug prices and the health and welfare of its population.

The strategic response of firms in the emerging countries like India has been positive. They began to compete and collaborate at the same time; they began to invest in R&D and innovate; and they began to engage heavily in offshore outsourcing projects for the developed-world's MNEs in drug discovery, clinical trials, process development, and other areas. In terms of providing the developing world with affordable drugs, the developed world's MNEs have several options, which include differential pricing, disguised price discrimination, out-licensing, sale of technologies, and marketing own authorized generics.

Solutions for providing drugs at affordable prices without disincentivising innovation go beyond the normal market mechanisms. They include close collaboration between the industry and the academic institutions in research, and a strong role for foundations that provide research funding to battle diseases afflicting the developing world. Public policy options include differential patent protection with the aim of increasing private sector incentives to discover drugs for diseases that predominately affect poor countries and expanded research funding aimed at neglected diseases.

Finally, it is clear that the traditional vertically integrated pharmaceutical MNE is being rapidly replaced by a network of businesses, academic institutions, governments, international organizations, and nonprofit foundations, customers, and suppliers. There is a need for greater understanding of the implications of the MNE–academic–public–

nonprofit research links. To the extent that the measured contribution of such links to the MNE profitability is large, it provides an *economic justification* for socially responsible behavior as well as a public policy framework for sustainable development.

KEY TERMS

differential patent protection
differential pricing
diguised price discrimination
generic firms
harmonization of IPRs
international price discrimination
private versus social rate of return on R&D
pull versus push approach to research funding
reference pricing
socially responsible pricing
TRIPs
WTO

DISCUSSION QUESTIONS

1. Describe key players and institutions in the global biopharmaceutical industry.
2. Discuss the main issues between the developed world and the developing world leading up to the TRIPs in the context of the global biopharmaceutical industry.
3. What are the pros and cons of TRIPs as it applies to the pharmaceutical industry?
4. What strategies would you recommend to the developed world's pharmaceutical MNEs to increase the appropriability of innovations worldwide?
5. What strategies would you recommend to emerging-country pharmaceutical firms in light of TRIPs enforcement?
6. What important public policy issues are raised—for developed and developing countries—by the application of TRIPs to the global pharmaceutical industry?

NOTE

This chapter is, in part, a revised and updated version of P.M. Rao, "The Emergence of the Pharmaceutical Industry in the Developing World and Its Implications for MNE Strategies," *International Journal of Pharmaceutical and Healthcare Marketing* 2 (2) (2008); and P.N. Ghauri and P.M. Rao, "Intellectual Property, Pharmaceutical MNEs and the Developing World," *Journal of World Business* 44 (2) (April 2009). The authors gratefully acknowledge the permission to reproduce parts of the above articles from Emerald and Elsevier, respectively.

REFERENCES

Arrow, K.J. 2004. New antimalarial drugs: Biology and economics meet. *Health and Development*, pp. 20–21. Washington, DC: International Monetary Fund.

Association of University of Technology Managers. 2010. U.S. Licensing Activity Survey, FY 2010. Available at http://www.technologytransfertactitics.com.

Barton, J.H. 2004. TRIPs and the global pharmaceutical market. *Health Affairs* 23 (3): 146–154.

Berndt, E.R. 2002. Pharmaceuticals in the U.S. health care: Determinants of quantity and price. *Journal of Economic Perspectives* 16 (4) (Fall): 45–66.

Bill & Melinda Gates Foundation. 2010. Progress and partnerships, Annual Report. At http://www.gatesfoundation.org/annualreport/2010/Documents/2010-annual-report-ceo-letter-english.pdf.

Bloom, D.E., D. Canning, and D.T. Jamison. 2004. Health, wealth and welfare. *Health and Development*. International Monetary Fund, pp. 10–15 (December). Washington, DC.

Bloom, D.E., D. Canning, and P. Malaney. 2000. Demographic change and economic growth in Asia. *Supplement to Population and Development Review* (26): 257–290.

Braga, C.A.P. 1995. Trade-related intellectual property issues: The Uruguay Round agreement and its economic implications. In *The Uruguay Round and the Developing Economies*, ed. W. Martin and L.A. Winters, pp. 381–411. Washington, DC: The World Bank.

Bruche, G. 2011. Emerging Indian pharma multinationals: Latecomer catch-up strategies in a globalized high tech industry. Berlin School of Economics and Law. Berlin, Germany.

Bull, P.F., and G.M. McEvoy. 1999. Creating and sustaining ethical capability in the multi-national corporation. *Journal of World Business* 34 (4) (Winter): 326–343.

Burstall, M., and J. Dunning. 1985. International investment in innovation. In *Pharmaceuticals among sunrise industries*, ed. N. Wells, pp. 185–197. New York: St. Martin's Press.

Chesbrough, H.W. 2003. The era of open innovation. *MIT Sloan Management Review* 44 (3) (Spring): 35–41.

Cockburn, I., and R. Henderson. 1996. Public-private interaction in pharmaceutical research. Proceedings of the National Academy of Sciences of the United States, 93: 12725–12730.

Cutler, D., A. Deaton, and A. Lleras-Muney. 2006. *Journal of Economic Perspectives* 20 (3) (Summer): 97–120.

Danzon, P.M. 1997. Price discrimination for pharmaceuticals: Welfare effects in the US and the EU. *International Journal of the Economics of Business* 4 (3): 301–321.

———. 2000. Pharmaceutical benefit management: An alternate approach. *Health Affairs* 19 (2): 24–30.

———. 2006. Economics of the pharmaceutical industry. *NBER Reporter* (Fall). Cambridge, MA: National Bureau of Economic Research.

DiMasi, J.A., and H.G. Grabowski. 2007. The cost of biopharmaceutical R&D: Is biotech different? *Managerial and Decision Economics* 28: 469–479.

DiMasi, J.A., R.W. Hansen, and H.G. Grabowski. 2003. The price of innovation: New estimates of drug development costs. *Journal of Health Economics* 22: 151–185.

Dorsey, E. Ray et al. 2010. Financial anatomy of biomedical research 2003–2008, *JAMA*, 303 (2): 137–143.

Friedman, Y. 2010. Location of pharmaceutical innovation: 2000–2009. *Nature Reviews Drug Discovery* 9: 835–836.

Gallini, N.T. 2002. The economics of patents: Lessons from recent U.S. patent reform. *Journal of Economic Perspectives* 16 (2): 131–154.

Ghauri, P.N., and P.M. Rao. 2009. Intellectual property, pharmaceutical MNEs and the developing world. *Journal of World Business* 44 (2) (April): 206–215.

Grabowski. H.G., J. Vernon, and J.A. DiMasi. 2002. Returns to research and development for the 1990s new drug introductions. *Pharmacoeconomics* 20: 11–29.

Hall, B.H. 2001. *The global nature of intellectual property: Discussion*. Toronto IP Conference. http://emlab.berkeley.edu/users/bhhall/bhpapers.html.

Hellerstein, R. 2003. *Do drug prices vary across rich and poor countries?* International Research, Federal Reserve Bank of New York.

International Federation of Pharmaceutical Manufacturers Association (IFPMA). 1995. *GATT TRIPs and the pharmaceutical industry: A review*, pp. 1–16. Geneva: IFPMA.

Joseph, R.K. 2011. The R&D scenario in Indian pharmaceutical industry. *Research and Information System for Developing Countries*. Discussion Paper no. 176 (December). New Delhi, India.

Kenney, M., and D. Patton. 2005. Entrepreneurial geographies: Support networks in three high-technology industries. *Economic Geography* 81 (2): 201–228.

Kollewe, J. 2012. Roche to offer discounted cancer drugs in India. *The Guardian*, March 23. At http://www.guardian.co.uk/business/2012/mar/23/roche-discounted-cancer-drugs-india.

Kremer, M. 2002. Pharmaceuticals and the developing world. *Journal of Economic Perspectives* 16 (4): 67–90.

Kremer, M., and R. Glennerster. 2004. *Strong medicine: Creating incentives for pharmaceutical research on neglected diseases*. Princeton, NJ: Princeton University Press.

Lanjouw, J.O. 1997. The introduction of pharmaceutical product patents in India: Heartless exploitation of the poor and suffering? NBER Working Paper No. 6366. Cambridge, MA: National Bureau of Economic Research.

———. 2003. Intellectual property and the availability of pharmaceuticals in poor countries. In *Innovation Policy and the Economy,* ed. A.B. Jaffe, J. Lerner, and S. Stern, pp. 91–130. National Bureau of Economic Research. Cambridge, MA: MIT Press.

———. 2005. Patents, price controls and access to new drugs: How policy affects global market entry. NBER Working Paper No. 11321. Cambridge, MA: National Bureau of Economic Research.

Leebron, D.W. 1996. Lying down with procrustes: An analysis of harmonization claims. In *Fair trade and harmonization*, ed. J. Bhagwati and R.E. Hudec, pp. 41–118. Cambridge, MA: MIT Press.

Lichtenberg, F.R. 2003. Pharmaceutical innovation, mortality reduction and economic growth. In *Measuring the gains from medical research*, ed. K.M. Murphy and R.H. Topel, pp. 74–109. Chicago: The University of Chicago Press.

———. 2010. Pharmaceutical price discrimination and social welfare. *Capitalism and Society* 5 (1): 1–29.

McNeil Jr., D.G. 2007. Drug partnership introduces cheap antimalarial pill. *New York Times*, March 1, A3.

Mehta, S. 2004. The emerging role of academia in commercializing innovation. *Nature Biotechnology* 22: 21–24.

Mitra, Barun. 2006. *Innovative intervention in the IP debate: Rise of the knowledge economy*. PowerPoint. New Delhi: Liberty Institute. http://www.stockholm-network.org/downloads/publications/ip/d41d8cd9-1P%20Academy%20-%20Module%202%20-%20Feb2006%20%-%Mitra.pdf.

Morel, Carlos et al. 2005. Health innovation in developing countries. *Innovation Strategy Today*, e-journal, 1 (1): 1–15.

Mowery, D.C., R.R. Nelson, B.N. Sampat, and A.A. Ziedonis. 2001. The growth of patenting and licensing by U.S. universities: An assessment of the effects of the Bayh-Dole Act of 1980. *Research Policy* 30: 99–119.

National Science Board. 2012. *Science and engineering indicators 2006*. Washington DC: U.S. Department of Commerce.

Organization for Economic Co-operation and Development (OECD). 2006. *Innovation in pharmaceutical biotechnology: Comparing national innovation systems at the sectoral level.* Paris: OECD.

Pharmaceutical Research and Manufacturers of America (PhRMA). 2010. *Pharmaceutical Industry Profile 2010*. Washington, DC: PhRMA.

———. 2011. *Pharmaceutical Industry Profile 2011*. Washington, DC: PhRMA.

Pisano, G. 2006a. Profiting from intellectual property revolution. *Research Policy* 35: 1122–1130.

———. 2006b. *Science business: The promise, the reality, and the future of biotech*. Boston, MA: Harvard Business School Press.

Pollack, A. 2003. UCSD getting more involved in drug trials. SignOnSanDiego.com, July 31.

Rao, P.M. 1998. Intellectual property protection and foreign direct investment: A global pharmaceutical industry perspective. In *Globalization, privatization and free market economy*, ed. C.P. Rao, pp. 232–253. Westport, CT: Quorum Books.

———. 2008. The emergence of the pharmaceutical industry in the developing world and its implications for MNE strategies. *International Journal of Pharmaceutical and Healthcare Marketing* 2 (2): 103–116.

Sampat, B.N. 2006. Patenting and U.S academic research in the 20th century: The world before and after Bayh–Dole. *Research Policy* 35: 772–789.

Sampath, Gehl. 2006a. India's product patent protection regime: Less or more of "Pills for the poor." Working Paper Series No. 2006-019. United Nations University, Maastricht Economic and Social Research and Training Centre on Innovation and Technology, 1–55.

———. 2006b. Economic aspects of access to medicines after 2005: Product patent protection and emerging firm strategies in the Indian pharmaceutical industry. United Nations University: Institute for New Technologies (INTECH): 1–106.

Scherer, F.M. 2004. The pharmaceutical industry: Prices and progress. *New England Journal of Medicine* 351(August 26): 927–932.

Stiglitz, J.E. 2003. *Globalization and its discontents*. New York: W.W. Norton.

Unnikrishnan, C.H. 2007. Pharma offshoring will present a $7 billion opportunity by 2013. Available at http://www.livemint.com/2007/04/30001103/pharma-offshoring-will-present.html.

Vachani, S., and N.C. Smith. 2004. Socially responsible pricing: Lessons from pricing of AIDS drugs in developing countries. *California Management Review* 47: 117–144.

Wilson, D. 2011. Facing generic Lipitor rivals, Pfizer battles to protect its cash cow. *New York Times*, November 30, p. B1.

World Health Organization (WHO). 2008. *The global burden of disease: 2004 update*. Geneva: WHO.

———. 2011. *Global HIV/AIDS response*. Progress Report, p. 24. Geneva: WHO.

———. 2012. *Global HIV/AIDS response*. Progress Report, Table 4, p. 99. Geneva: WHO.

World Trade Organization. 2001. Ministerial Declaration. WT/MIN(01)DEC/1 (20 November). Available at http://www.wto.org/english/thewto_e/minist_e/min01_e/mindecl_e.htm.

Yamin, M., and J. Otto. 2004. Patterns of knowledge flows and MNE innovative performance. *Journal of International Management* 10 (April): 239–258.

Index

About the Authors

P.M. Rao

P.M. Rao is a professor of marketing and international business at the College of Management, LIU, Post. He received his PhD in economics from New York University Leonard N. Stern School of Business. Dr. Rao has developed and taught a variety of graduate and undergraduate courses at LIU Post; in France and Switzerland in the accelerated MBA program; and in corporate programs at Motorola, Grumman, and Verizon. These courses include Marketing High-Technology in a Global Business Environment, Product Strategy, Global Business, and Competing in the Global Trading System.

Prior to joining the academy, Dr. Rao worked for eighteen years at AT&T in various managerial positions involving studies and analyses on the economics of innovation, demand and cost models, antitrust, organization and funding of Bell Labs, productivity and strategic planning.

He is an active researcher and has published extensively in areas that include strategies for high-technology firms, intellectual property issues in software and pharmaceutical industries, and internationalization of multinational R&D. Dr. Rao has presented papers at such international conferences as the Academy of International Business, International Telecommunications Society, Academy of Marketing Science, and the Association of Marketing Theory and Practice. He has published numerous articles in scholarly journals, including the *Journal of World Business, International Business Review*, *Competitiveness Review*, *Journal of Quantitative Economics*, *Journal of Marketing Theory and Practice*, and the *Journal of High-Technology Management*. Dr. Rao has delivered public lectures at the Pôle University Léonard de Vinci in Paris and at the Curtin University of Technology in Australia for academic, business, and student audiences.

JOSEPH A. KLEIN

Joseph A. Klein is an attorney with a Harvard Law School degree who specializes in information technology law, commercial contracts, and antitrust law. He is currently a partner with the law firm FisherBroyles LLP. Prior to joining FisherBroyles, Mr. Klein worked for more than three decades in senior counsel positions at AT&T, Bell Communications Research, and Merrill Lynch. As a result, he has extensive experience in the negotiation and drafting of sophisticated commercial telecommunications, outsourcing software development, application hosting, technology consulting, and information license agreements. Mr. Klein's antitrust work has focused on joint ventures, standards setting, and intellectual property licensing practices.

Complementing his hands-on, real-world experience with complex high-tech commercial transactions, Mr. Klein has authored or co-authored several articles on the subjects of legal protections for intellectual property, technology licensing and transfer issues, privacy, and antitrust.

For Product Safety Concerns and Information please contact our EU representative GPSR@taylorandfrancis.com
Taylor & Francis Verlag GmbH, Kaufingerstraße 24, 80331 München, Germany

www.ingramcontent.com/pod-product-compliance
Lightning Source LLC
LaVergne TN
LVHW080651011026
849160LV00014B/1384